D0209048

More Surprises in
Theoretical Physics

Princeton Series in Physics

Edited by Philip W. Anderson, Arthur S. Wightman, and Sam B. Treiman

Quantum Mechanics for Hamiltonians Defined as Quadratic Forms *by Barry Simon*

Lectures on Current Algebra and Its Applications *by Sam B. Treiman, Roman Jackiw, and David J. Gross*

Physical Cosmology *by P.J.E. Peebles*

The Many-Worlds Interpretation of Quantum Mechanics *edited by B. S. DeWitt and N. Graham*

Homogeneous Relativistic Cosmologies *by Michael P. Ryan, Jr., and Lawrence C. Shepley*

The $P(\phi)_2$ Euclidean (Quantum) Field Theory *by Barry Simon*

Studies in Mathematical Physics: Essays in Honor of Valentine Bargmann *edited by Elliott H. Lieb, B. Simon, and A. S. Wightman*

Convexity in the Theory of Lattice Gases *by Robert B. Israel*

Works on the Foundations of Statistical Physics *by N. S. Krylov*

Surprises in Theoretical Physics *by Rudolf Peierls*

The Large-Scale Structure of the Universe *by P.J.E. Peebles*

Statistical Physics and the Atomic Theory of Matter, From Boyle and Newton to Landau and Onsager *by Stephen G. Brush*

Quantum Theory and Measurement *edited by John Archibald Wheeler and Wojciech Hubert Zurek*

Current Algebra and Anomalies *by Sam B. Treiman, Roman Jackiw, Bruno Zumino, and Edward Witten*

Quantum Fluctuations *by E. Nelson*

Spin Glasses and Other Frustrated Systems *by Debashish Chowdhury* (*Spin Glasses and Other Frustrated Systems* is published in co-operation with World Scientific Publishing Co. Pte. Ltd., Singapore)

Large-Scale Motions in the Universe: A Vatican Study Week *edited by Vera C. Rubin and George V. Coyne, S.J.*

Instabilities and Fronts in Extended Systems *by Pierre Collet and Jean-Pierre Eckmann*

More Surprises in Theoretical Physics *by Rudolf Peierls*

RUDOLF PEIERLS

More Surprises in Theoretical Physics

PRINCETON UNIVERSITY PRESS
PRINCETON · NEW JERSEY

Published by Princeton University Press, 41 William Street,
Princeton, New Jersey 08540
In the United Kingdom: Princeton University Press, Oxford

Library of Congress Cataloging-in-Publication Data

Peierls, Rudolf Ernst, Sir, 1907–
More surprises in theoretical physics / Rudolf Peierls.
p. cm. — (Princeton series in physics)
Includes bibliographical references.
ISBN 0-691-08576-5 (cl : alk. paper)
ISBN 0-691-02522-3 (pbk. : alk. paper)
1. Mathematical physics. I. Title. II. Series.
QC20.P345 1991 530.1'5—dc20 90-23189

This book has been composed in Linotron Trump

Princeton University Press books are printed on acid-free paper,
and meet the guidelines for permanence and durability of the
Committee on Production Guidelines for Book Longevity of the
Council on Library Resources

Printed in the United States of America by Princeton University Press,
Princeton, New Jersey

10 9 8 7 6 5 4 3 2 1

10 9 8 7 6 5 4 3 2 1 (Pbk.)

Designed by Laury A. Egan

Contents

Preface

The kind reception of my *Surprises in Theoretical Physics* (1979) persuaded me to look for more surprises, and this little volume is the result. As before, I have tried to collect examples in which the outcome of a calculation is different from what one would have guessed originally, or cases in which the solution turns out to be much simpler, or much harder, than one at first imagined. In all cases it is clear that there would have been no surprise if one had really understood the problem from the start.

As before, I have selected mainly surprises that have arisen in my own work or that of my collaborators (which should explain the predominance of my name in the list of references), or which I have found exciting. They tend to be concerned with rather old problems. This does not mean that I have encountered no surprises more recently, but that the recent ones were not simple enough to be presented in the direct form I have found most appropriate.

Many of the problems and some of the answers will be familiar to the experts in the relevant fields, but I have tried to make them intelligible and interesting to those working in other fields as well.

I am grateful to Ian Aitchison, David Brink, Robin Stinchcombe, and Marshall Stoneham for constructive criticism of some sections. The text owes much to the careful and efficient editing by Alice Calaprice.

**More Surprises in
Theoretical Physics**

1

General Quantum Mechanics

1.1 Some Simple Arguments in Wave Mechanics

Any theoretical physicist has met, in his introduction to the subject, the simplest examples of Schrödinger's equation, including the harmonic oscillator. In demonstrating its solution, it is usually shown that for energies satisfying the usual quantum condition,

$$E = (n + \tfrac{1}{2})\hbar\omega, \tag{1.1.1}$$

where n is a non-negative integer and ω the frequency, the equation has a solution satisfying the correct boundary conditions. It is equally important to know that these are the only solutions, i.e., that for an energy not equal to (1.1.1) no admissible solution exists. This negative statement is not usually proved in elementary treatments, or else it is deduced from quite elaborate discussions of the convergence and behavior of a certain infinite series. It is therefore surprising to find that the result can be seen without any complicated algebra.

We recall that the equation can, in suitable units, be written as

$$\frac{d^2\psi}{dx^2} + \left(E - \frac{1}{2}x^2\right)\psi = 0, \tag{1.1.2}$$

the unit of length being $\hbar/m\omega$, that of energy being $\hbar\omega/2$.

By including a factor that represents the dominant behavior at infinity,

$$\psi(x) = v(x).\exp\left(-x^2/2\right), \tag{1.1.3}$$

we can expand $v(x)$ in a power series, and in a familiar way get a two-term recurrence formula for the coefficients. The condition that the series should terminate, i.e., that v be a polynomial of degree n, is precisely (1.1.1). Such a polynomial, multiplied by the exponential in (1.1.3), vanishes at infinity and is an admissible solution.

It is easy to show there are no other solutions if we remember the theorem that the $(n + 1)$th solution of the Schrödinger equation (the ground state being $n = 0$) has n nodes. This theorem becomes evident if one considers the graph of $\psi(x)$ as always starting with $\psi(-\infty) = 0$, and raising the energy gradually. If the energy is below the potential minimum, $(E - V)$ is everywhere negative, so the curve for ψ is everywhere curved away from the x axis. As it has started rising, it can never come back to the axis. If E is increased, there appears a region of curvature toward the axis, and the size of that region and the curvature in it increase until the graph manages to bend back and reach the axis again. This is the ground state, and it has no node.

If the energy is raised further, ψ crosses the axis, and only after a further rise in energy will it bend back in its second loop and hit the axis again. We have now reached the second state, with one node. This process continues, leading to the stated theorem.

Now the solution we have constructed for the harmonic oscillator contains, apart from the exponential factor, which does not vanish, a polynomial of degree n, which has at most n nodes, and by the theorem can be *at most* the $(n + 1)$th solution. But as we have constructed solutions for any n, there are n solutions of lower energy, so it must be *at least* the $(n + 1)$th. It follows that it must be exactly the $(n + 1)$th, and there are no others.

Another step that is usually believed to be very hard to prove is the completeness of the eigenstates of the Schrödinger equation, i.e., the fact that any "reasonable" function f of the coordinates can be expanded in a series of the form

$$f = \sum_n a_n \psi_n. \tag{1.1.4}$$

It is easy to show that if f can be expressed in the form (1.1.4), then, subject to some conditions of convergence, the coefficients must have the values

$$a_n = \int \psi_n^* f dv, \tag{1.1.5}$$

where dv is the volume element in coordinate space.

Now assume there exists a particular function f that cannot be so expanded. We can still define coefficients by (1.1.5) and insert them in the series (1.1.4). If this series converges, it represents a function, say g, which by hypothesis is not identical with f.

As regards the convergence, let us simplify the problem by assuming that the potential energy is analytic and has a finite upper bound. Then, for energies large compared to this, the coefficients (1.1.5) asymptotically approach the Fourier components of f, and the series must converge uniformly if f has a uniformly convergent Fourier series. This is certainly the case if f is everywhere continuous and has a continuous derivative.

Now consider the function $f - g$, which by hypothesis is not identically zero. It is also orthogonal to all the ψ_n, since the projection of f on ψ_n is again (1.1.5). (Note that the uniform convergence of the series for g allows the order of summation and integration to be interchanged.)

Now consider the class of functions that are orthogonal to all the ψ_n, and find among this class the function that gives the least value of the quantity,

$$I = \int dv\, \{(-\hbar^2/2m)|\nabla\phi|^2 + V\phi^2\}/|\phi|^2\, dv. \qquad (1.1.6)$$

The Euler equation of the variation principle of I is the Schrödinger equation; the function that minimizes I is the eigenfunction of the ground state. If we require the function that minimizes I among the class of functions that are orthogonal to a certain number, say N, of the eigenfunctions, we must add terms with Lagrange multipliers:

$$\delta\{I + \Sigma\mu_n 2Re \int dv\, \psi_n^* \phi\} = 0. \qquad (1.1.7)$$

Carrying out the variation and setting the coefficient of $\delta\phi^*$ equal to zero (using the fact that the real and imaginary parts of ϕ are independent, which amounts algebraically to treating ϕ and ϕ^* as independent), we find

$$\{(-\hbar^2/2m)\nabla^2 + V - I\}\phi + \Sigma\mu_n\psi_n = 0. \qquad (1.1.8)$$

Multiply this equation by a particular ψ_n^* and integrate. After integrating the first term by parts we find

$$(E_n - I)\int \psi_n^* \phi\, dv + \mu_n = 0. \qquad (1.1.9)$$

But the integral vanishes, since ϕ is by definition orthogonal to ψ_n. Therefore the right choice of the Lagrange parameter μ_n is zero, making the solution by (1.1.8) an eigenfunction of the Schrödinger equation with eigenvalue I. This procedure is applicable with $N = \infty$, making the required function orthogonal to *all* eigenfunctions. Since we found it to be itself an eigenfunction, this is impossible, and therefore the assumption that it exists is disproved.

Note: A careful mathematician might have noticed that the class of functions we have defined is not a closed set, so that the limit of a sequence of functions within that class might not belong in it, and this would suggest that the minimum might not be reached within the class. However, the only way a sequence can lead out of the class is by converging to a function with a discontinuous derivative. By rounding off the incipient corner of these functions, one can reduce the value of I for each of them, and therefore also for the limiting function, which cannot be a candidate for minimum I.

The simplifying restrictions, to an analytic potential energy with a finite upper bound, and to functions with a continuous first derivative, can be widened without using much more sophisticated mathematics than has been used above.

1.2 Observations and the "Collapse of the Wave Function"

In section 1.6 of *Surprises in Theoretical Physics* (Princeton, 1979; hereafter referred to as *Surprises*), I have discussed the interpretation of quantum mechanics and emphasized that there is no alternative to the standard view that the wave function (or more generally the density matrix) represents our knowledge of a physical system. I also briefly referred to the "contraction" (or "collapse") of the wave function. I find there is more to be said on this subject.

Let us first recall that cases in which we are dealing with a wave function (sometimes called "pure states") are academic idealizations, and that normally we have to use a density matrix. This, first introduced independently by Landau and von Neumann, and described more clearly by Dirac, is appropriate when we are concerned with a "mixture," i.e., a combination of different quantum states without phase relations, so that their probabilities, not their amplitudes, have to be added together.

Such a density matrix has to be used whenever our observations are not carried out to the limit set by the uncertainty principle, in other words, when there are quantities we could have measured without disturbing the knowledge already acquired, but which have not been measured. This kind of lack of knowledge is the only kind possible in classical physics, and so a density matrix may be said to contain both quantum mechanical and classical kinds of ignorance.

The formal expression for the density matrix is, for example, in coordinate (Schrödinger) representation,

$$(x|\rho|x') = \sum_n \rho_n \, \psi_n(x) \, \psi_n^*(x'), \tag{1.2.1}$$

where the ψ_n are the eigenstates of a full set of observables, including those quantities we could have, but did not, observe, and $a\rho_n$ the probability that the observation would have yielded the result n.

From this density matrix the expectation value of any quantity A can be obtained by

$$\langle A \rangle = \sum \rho_n \langle n|A|n\rangle = \int dx \, dx' \, \langle x|A|x'\rangle\langle x'|\rho|x\rangle = Tr(A\rho). \tag{1.2.2}$$

The probabilities must add up to unity, which implies

$$Tr(\rho) = 1. \tag{1.2.3}$$

A pure state is a special case of the density matrix, in which, in some representation, only one of the weights ρ_n in (1.2.1) is equal to unity and the others zero. Without reference to a special representation, this can be expressed by the condition

$$\rho^2 = \rho. \tag{1.2.4}$$

If we carry out an observation, for example if we observe the spin component of an atom of spin ½, we have to replace our previous wave function (or density matrix) by an eigenfunction of the spin. This is the "collapse of the wave function," which we want to examine in some more detail.

Note that in this way of talking, our "system" is just the atom and its spin; we do not specify how we observe it; we assume that we have a way of finding out what its spin component is. We can extend our definition of the system by including in the quantum mechanical description some part of the apparatus used for the observation, but there will always be an observer outside the system.

It is sometimes thought that the "collapse" is the result of the interaction of the system in question with the measuring apparatus. This can be elucidated by including some of the measuring instruments in the quantum mechanical description.

In *Surprises* I discussed a Stern-Gerlach experiment as an example of an observation of spin. In the inhomogeneous field of the Stern-Gerlach magnet, the atom is deflected one way or another, according to whether the appropriate component of its spin, say s_z, is $+\frac{1}{2}$ or $-\frac{1}{2}$. In other words, the effect of the magnet is to set up a correlation between the spin component and the position of the atom, which we are now including in the description; but we still need the observer, who "sees" which way the atom has been deflected. Failing this we have not yet obtained any knowledge of the spin; either direction is still as probable as it was before. This goes together with the fact that we have not yet lost irretrievably any knowledge we might have had of another component, say s_x, which, by the uncertainty principle, would be incompatible with knowledge of s_z.

Indeed it would be possible, in principle, to deflect the two beams emerging from the magnet, so as to recombine them, and to compensate the precession that would have altered s_x on the way. This would restore the old value of s_x so we would have retrieved our original knowledge, while of course we would have spoiled the measurement of s_z.

The act of the observer seeing the position of the atom must therefore make this retrieval impossible. To see how this happens, we may further extend the "system" included in our description. We therefore include the next step, which is the device used to detect the position of the atom. This could, for example, be a counter in one of the beams. We now have a triple correlation, because there are now the two alternatives: (a) spin up, atom in upper beam, counter activated; and (b) spin down, atom in lower beam, counter not activated. But each of these alternatives still has the same probability as before the "measurement"; without observing the state of the counter we have not yet acquired new information. Have we at least managed to lose the knowledge of s_x?

At first sight this seems plausible. If we idealize the counter to have only two states, u_o being the wave function of the normal state, and u_1 that of the activated counter, the wave function of the whole system after the passage of the atom will be

$$A\psi_+(r)u_1 + B\psi_-(r)u_0, \tag{1.2.5}$$

where $\psi_+(r)$ and $\psi_-(r)$ are wave functions for an atom with either value of s_z, with unit amplitude, and A, B are amplitude factors. The information on s_x depends on the phase relation between the first and second term of (1.2.5), i.e., between A and B. To find their relative phases by experiment, we would have to observe an operator, say Q, and the expectation value of Q would be

$$\begin{aligned}\langle Q\rangle = \int dr\, dr'\{&(r|Q_{11}|r')A^*A\psi_+^*(r)\psi_+(r')\\ &+ (r|Q_{10}|r')A^*B\psi_+^*(r)\psi_-(r')\\ &+ (r|Q_{01}|r')B^*A\psi_-^*(r)\psi_+(r')\\ &+ (r|Q_{00}|r')B^*B\psi_-^*(r)\psi_-(r')\}.\end{aligned} \tag{1.2.6}$$

To make this sensitive to the relative phase of A and B, we must not only make the two wave functions of the atom overlap, but the cross matrix element Q_{10} must be non-zero, in other words, our observation must be capable of activating or deactivating the counter coherently. While this is possible in principle, it would be complicated in practice.

An actual counter has many more degrees of freedom, not all of which it is practical to determine fully to the limits of the uncertainty principle. Instead of the two wave functions for the counter, we now have to use density matrices, which will obscure the coherence between the two terms and make the restoration of the lost information even more difficult.

Again no information is gained, and the unwanted information is not irretrievably lost (though hard to retrieve) until the observer "sees" the state of the counter.

To do so, the observer may have arranged the counter to be part of an electric circuit containing some recorder. If we include this in the "system" to be described, there are further correlations, and the retrieval of the unwanted information becomes more difficult. However, the probability of the z component of the spin being $+\frac{1}{2}$ or $-\frac{1}{2}$ remains unchanged, until we "see" the recorder.

The general pattern is now clear: there must always be a division between the system described by quantum mechanics and the observer with his apparatus. There is great freedom in placing this division. We may include in the "system" only the atomic spin and say we see the z component of the spin positive (or negative) and change our wave function accordingly. We may follow it all the way as just discussed and place the division after the recorder, saying: we see from the recorder that the counter has operated and therefore that the atom was in the upper beam, and that the spin was up (or vice versa). As far as the atom is concerned, the results are the same, as they would also be for any intermediate placing of the division. The ease of changing the division is due to the fact that all but the first steps in the chain are to a good approximation classical, so that we can discuss their correlations in classical terms.

As was discussed in *Surprises*, we could push this division farther and farther back, until it gets to our brain and we get involved with the concept of "knowledge" or consciousness.

To sum up, a measurement must contain three elements: (1) generating a correlation between the quantity to be measured and some variable in the measuring apparatus; (2) an interference with the variables that the uncertainty principle says are not compatible with the knowledge to be obtained; and (3) a recognition (not contained in the quantum mechanical description) which of the possible results of the measurement is observed.

The first two steps are necessary but not sufficient. Indeed, it is possible to make a "bad" measurement, which interferes with the state of the system but does not produce any new information.

We can now give a more precise answer to the question, "Whose knowledge does the density matrix describe?" It is possible for two observers to have some knowledge of the same system, and the knowledge possessed by one may differ from the other. For example, one may have only an incomplete or inaccurate view of the measuring instrument. In this situation the two observers will use different density matrices.

However, the information possessed by the two observers must not contradict the uncertainty principle. For example, if one observer knows the x component of an atomic spin, the other may not know the z component. This is because the measurement made by the first observer would have caused an uncontrollable interference with the z component. This limitation can be expressed concisely by saying that the density matrices appropriate to the two observers must commute with each other.

This can be seen by considering the representation in which one of these matrices is diagonal. If the other does not commute with it, it must have off-diagonal elements in the representation. This means that it shows phase relations between the states which the first observer could have observed, and this would violate the uncertainty principle.

At the same time, the two observers should not contradict each other. This means the product of the two density matrices should not be zero.

Indeed, take a representation in which both are diagonal (which is possible if they commute); then there must be at least some states for which the probabilities assumed by both observers are non-zero, and that means the product is non-zero.

The need to refer to an observer is related to the fact that the tools of the quantum mechanical description, i.e., the wave function or density matrix, represent, as I have stressed, the observer's knowledge of the system.

The following question is sometimes raised: We are trying to apply quantum mechanics to the early stages of the universe when no observers were present. Is that not a contradiction? The answer is that the observer does not have to be present at the time of the event. We can derive knowledge of the early universe from present-day evidence, such as the cosmic microwave radiation, and in this sense we are observers. If there are events beyond the cosmic horizon, it makes no sense to apply quantum mechanics to them.

All these statements are rather straightforward consequences of the principles of quantum mechanics (though the precise formulation is a matter of choice); the surprise is that so many people still have difficulty understanding this.

1.3 Application of the W.K.B. Method in Several Dimensions

The "W.K.B. method" (first used by Harold Jeffreys) is explained in every textbook on quantum mechanics. But the books usually deal only with its application to one-dimensional problems, and that is not surprising because the method is not powerful enough to give an approximation to the wave function in more than one dimension.

However, there are important cases, particularly relating to tunneling through potential barriers, in which the method will at least yield an upper bound to the transmission probability. We shall look at the equations for the method to see both the nature of the difficulty and what can be done.

Write the wave function of a particle as

$$\psi(\mathbf{r}) = \exp\{-S(\mathbf{r})/\hbar\}. \tag{1.3.1}$$

Then the Schrödinger equation becomes

$$(\nabla S)^2 - \hbar\nabla^2 S = 2m\,(V - E), \tag{1.3.2}$$

E being the energy and V the potential.

The W.K.B. approximation is intended for the conditions of geometrical optics, in which the gradient of S is large, but slowly variable. This means that the second term on the left-hand side of (1.3.2) is small compared to the first. (This is suggested also by it containing the factor $\hbar$, since we are dealing with a semiclassical approximation, in which $\hbar$ is taken as small.) So in the W.K.B approximation,

$$(\nabla \mathbf{S})^2 = 2m(V - E). \tag{1.3.3}$$

In one dimension this equation determines the derivative of S, apart from the sign, and this makes it possible to find S by integration. It is well known that care has to be taken near the classical turning points, where $V - E = 0$, since there (1.3.3) is not an adequate approximation.

However, in more than one dimension, the gradient of S is a vector, and the equation determines only its magnitude and not its direction. We can still make use of the approximation, typically in a tunneling situation. In that case there is a region in which $V>E$, and we would like to find the probability of the particle reaching a point P in or beyond that region.

The function S will in general be complex. Writing

$$S = \sigma + i\tau, \tag{1.3.4}$$

we find

$$(\nabla \sigma)^2 - (\nabla \tau)^2 = 2m(V - E) \tag{1.3.5}$$

$$\nabla\sigma . \nabla\tau = 0. \tag{1.3.6}$$

We are interested in the probability density,

$$|\psi|^2 = \exp - 2\sigma, \tag{1.3.7}$$

and therefore in σ. We obtain an inequality for σ alone, by noting that the second term in (1.3.5) is negative:

$$|\nabla \sigma|^2 \geqslant 2m(V - E). \tag{1.3.8}$$

To find some information about σ we use the fact that

$$\sigma(P) = \sigma(A) + \int_A^P |\nabla\sigma|\, ds, \tag{1.3.9}$$

where the integral is taken along a line of steepest gradient of σ, and A is a point on that line. In general this line will lead into the region outside the potential barrier, where the wave function has an appreciable magnitude, and we choose the point A at the border of that region, i.e., where $V - E = 0$.

Normally there is only one such line through P, and we do not know it. The integral (1.3.9) is, however, greater than the minimum value of the integral, which in turn by (1.3.8) is greater than the integral of $\sqrt{\{2m(V - E)\}}$. Finally, we also do not know the point A, because there will be only one such point on the boundary lying on the line of steepest gradient. Again we obtain a lower bound on σ by choosing the point A' on the boundary for which our estimate of the integral is lowest.

We have now obtained the inequality

$$\sigma(P) - \sigma(A) \geqslant \text{Min} \int \sqrt{\{2m(V - E)\}}ds, \tag{1.3.10}$$

where the minimum is selected among all paths joining P to any point on the surface $V = E$.

From the way in which this inequality has been derived it might appear that it gives a very generous lower bound on σ, and thus a generous upper bound on the transmission probability. This is by no means always the case. There are cases in which one knows that in the relevant region S is real, so that (1.3.8) becomes an equality. Moreover, the equation for the path that makes the integral in (1.3.9) a minimum, for fixed A, is also the equation for the line of steepest gradient. There is only one line of steepest gradient passing through P, whereas there is a solution of the minimum equation for every point A. In general, we do not know the position of A'. However, if we know that the magnitude of the wave function had only one maximum in the neighborhood of the boundary, the line of steepest gradient must reach that maximum, so we then know the approximate position of A'.

The surprise, in this instance, is how much information can be obtained in spite of the difficulty stated at the beginning. The method reported here was first proposed by Kapur and Peierls (1937). It has been rediscovered independently by several authors. An example in which the inequality becomes an equality was discussed by P. L. Kapur (1937).

1.4 Perturbation Theory for Projected States

Projected states are an important auxiliary tool in some calculations. They are useful when a problem involving a symmetry is similar to a soluble problem without that symmetry. A typical example is a two-center, one-electron problem, as in the molecular ion H_2^+. Its Hamiltonian (taking the one-dimensional case for simplicity) is

$$H = T + V(x - a) + V(x + a), \tag{1.4.1}$$

where T is the kinetic energy and V the interaction potential of the electron with either of the centers, which are $2a$ apart. H obviously admits the symmetry transformation

$$x \rightarrow -x. \tag{1.4.2}$$

The eigenstates of H are therefore symmetric or antisymmetric in x. They must therefore satisfy the condition,

$$N\,\phi(x) \equiv \tfrac{1}{2}[\phi(x) + \gamma\,\phi(-x)] = \phi(x), \qquad (1.4.3)$$

where $\gamma = \pm 1$. In general there exists such a symmetry operator N, which as a projection operator has the property that $N^2 = N$, and which we shall assume Hermitian (although there are exceptions to this).

If the two centers are reasonably far apart, the motion of the electron will most of the time resemble that under the influence of the single center. There is therefore some similarity between the lowest eigenstates of the "molecule" and that of the "atom," which has the simpler Hamiltonian,

$$H_s = T + V(x - a), \qquad (1.4.4)$$

whose lowest eigenfunction ϕ_0 we may assume known.

However, this does not possess the required symmetry, which makes it physically very different from the correct eigenfunction of H. We can use it to construct a symmetric function,

$$\psi_0(x) = N\phi_0(x), \qquad (1.4.5)$$

using the symmetrizing projection operator N.

This projected function is for many purposes a reasonable approximation to the lowest state of given symmetry. It was used, for example, in the classical paper by Heitler and London on the H_2 molecule, which essentially uses this function as a trial function in the variation principle. But sometimes it is necessary to correct the error involved in this approximation, or at least to estimate the magnitude of the error. This suggests constructing a perturbation theory, which usually involves the use of a complete set of approximate solutions. A set that suggests itself is obtained by projecting not only the ground state, but all eigenfunctions of the "simple" (in our example the atomic) problem:

$$\psi_n = N\phi_n. \qquad (1.4.6)$$

However, there are two related difficulties: (a) The functions ψ_n are not the solutions of any simple Hamiltonian, since they have been obtained by mixing solutions of different simple Hamiltonians; we cannot therefore define a perturbing potential, which would be the difference between the complete and the approximate Hamiltonian. (b) Because of (a) these functions are not only not orthogonal, they are also overcomplete. This follows from the fact that they all have the same symmetry, and therefore evidently can be used to expand only functions of the same symmetry. They therefore span only part of function space, but their number is still the same as that of the complete set. (This counting of infinities is of course not rigorous, but it gives an intuitive justification for a statement that we shall prove.)

In other words, if we express some function f as a series of the ψ_n,

$$f = \Sigma\, a_n\, \psi_n, \tag{1.4.7}$$

this is possible only if f has the required symmetry, i.e., if

$$Nf = f, \tag{1.4.8}$$

and if this condition is satisfied there exist an infinite number of ways in which to choose the coefficients a_n. To see this, consider an expansion of f in terms of the ϕ_n, which do form a complete orthonormal set, so that a valid expansion is

$$f = \Sigma\, a_n \phi_n, \tag{1.4.9}$$

with

$$a_n = \langle \phi_n, f \rangle. \tag{1.4.10}$$

If we now operate on (1.4.9) with N, the left-hand side does not change, because of (1.4.8), and on the right the ϕ_n become ψ_n by (1.4.6). The equation therefore becomes (1.4.7) and the a_n defined by (1.4.10) are *one possible* set of expansion coefficients for f.

We can also expand in terms of the ϕ a function g which has a different symmetry, so that it is annihilated by the projection operator N:

$$Ng = 0. \tag{1.4.11}$$

We write

$$g = \Sigma\, b_n \phi_n; \tag{1.4.12}$$

$$b_n = \langle \phi_n, g \rangle. \tag{1.4.13}$$

Operating with N on (1.4.12) we see that

$$\Sigma b_n\, \psi_n = 0. \tag{1.4.14}$$

We see therefore that we can add the b_n to the a_n, without altering the sum. Since g can be an arbitrary function of "wrong" symmetry, we see that there is great freedom of choice in the coefficients in the series (1.4.7). One way in which we may try to construct an "unperturbed" Hamiltonian is to associate with the nth term in (1.4.7) the eigenvalue E_n of the soluble Hamiltonian H_s. But to do so unambiguously we must choose, for any given f, a particular set of expansion coefficients a_n. We have seen that a possible simple choice is to use (1.4.10). This is by no means the only acceptable choice, but it is the simplest one.

Once the choice is made it is straightforward to set up the perturbation theory. We write the wave function and the energy eigenvalue in the form

$$\psi = \psi_{p0} + u_1 + u_2 + \ldots \tag{1.4.15}$$

$$E = E_0 + \epsilon_1 + \epsilon_2 + \ldots \tag{1.4.16}$$

in increasing order of smallness (how small can be decided only by inspecting the results). We are starting from the ground state, ψ_0, for simplicity of notation, but we could have used any other unperturbed level instead.

Writing now $u_n = \Sigma c_n \psi_n$ and inserting in the Schrödinger equation, we find, to first order,

$$\Sigma c_n (H - E)\psi_n = (E - H)\psi_0. \tag{1.4.17}$$

The coefficients on the left are small of first order, so we need to know the effect of H only to zero order, and to that order we had agreed to associate with ψ_n the approximate eigenvalue E_n. Similarly we can on the left-hand side replace E by E_0. In other words, we can write for the left-hand side

$$\Sigma c_n (E_n - E_0)\psi_n. \tag{1.4.18}$$

To find the coefficients we must expand the right-hand side in a series of ψ_n and compare terms. By the convention we have chosen, this is done by projecting with the ϕ_n on the right-hand side. We find

$$c_n = \langle \phi_n | E - H | \psi_0 \rangle / E_n - E_0. \tag{1.4.19}$$

The numerator can be written

$$\langle \psi_n | E - H | \phi_0 \rangle. \tag{1.4.20}$$

We have used the fact that ϕ_0 satisfies (1.4.8) and that N commutes with the Hamiltonian H.

To keep the coefficient c_0 finite, the numerator must vanish for $n = 0$, which means that to first order the energy is given by

$$E = E_0 + \epsilon_1 = \langle\psi_0|H - E|\phi_0\rangle/\langle\psi_0|\phi_0\rangle. \qquad (1.4.21)$$

This is the value obtained by using ψ_0 as a trial function in the variation principle, which is as it should be. To make the approximation manifest, we can also write the numerator of (1.4.19) in the form

$$\langle\psi_n|E - H|\phi_0\rangle = \langle\psi_n|\epsilon + (H - H_s)|\phi_0\rangle, \qquad (1.4.22)$$

where we have used the fact that ϕ_0 is an eigenfunction of H_s with eigenvalue E_0.

$H - H_s$ acting on the unprojected ground state will be small in the circumstances we have envisaged. In our molecular example it represents the action of the potential of one center on the wave function of the electron under the influence of the other center, and that is small if the distance a is greater than the extension of the atomic eigenfunction.

The rest of the calculation proceeds like ordinary perturbation theory. For example, the second-order contribution to the energy is

$$\epsilon_2 = \sum_{n \neq 0} \frac{1}{\langle\psi_0|\psi_0\rangle} \frac{1}{E_0 - E_n} \langle\phi_0|H_s - \epsilon_1|\phi_n\rangle \langle\psi_n|H_s - \epsilon_1|\phi_0\rangle. \qquad (1.4.23)$$

The method outlined here was used for molecular problems as early as in the paper by Eisenschitz and London (1930) on van der Waals forces. They used the choice (1.4.10) without mentioning the ambiguity. It has since been used extensively in molecular calculations (see, e.g., Hirschfelder 1967; Amos 1970). The application to projected states in general was probably first suggested by me and developed by a number of others, in particular Ady Mann and Bülent Atalay.

The surprise is that although the type of problem considered here does not meet the conditions usually regarded as necessary for perturbation theory, there is a simple way out.

2

Condensed Matter

2.1 Brillouin Zones

In the early days of the electron theory of metals, Bloch had shown that the eigenfunctions of an electron in a perfectly periodic potential could be written in the form

$$\psi_{k,n}(\mathbf{r}) = u_{k,n}(\mathbf{r}).\ \exp i\mathbf{kr}. \tag{2.1.1}$$

Here u is a function with the same periodicity as the potential. In other words, u repeats in every unit cell of the crystal, and the variation from cell to cell of the wave function is given by the exponential factor. For each value of the wave vector $\mathbf{k}$ there is an infinite set of eigenfunctions with different u; they are labeled by n. The wave vector $\mathbf{k}$ ranges over a polyhedron known as the *first Brillouin zone,* which is defined in the following way: Consider the reciprocal lattice, consisting of all points $\mathbf{g}$ in $\mathbf{k}$ space, for which the factor exp $i\mathbf{gr}$ has the same periodicity as the lattice. Then construct the planes bisecting the line from the origin to any reciprocal lattice point $\mathbf{g}$. The space around the origin free of all these planes is the first Brillouin zone. For a simple square lattice in two dimensions, the construction is shown in figure 2.1. In one dimension the first zone is the interval $-\pi/a < k < \pi/a$, where a is the spacing.

The energy belonging to the wave function (2.1.1) is

$$E_n(\mathbf{k}). \tag{2.1.2}$$

This is evidently a continuous function of $\mathbf{k}$; the label n distinguishes continuous energy ranges known as *bands*. Bloch illustrated these results for the case of tightly bound electrons, for which near each atom the wave function is very nearly identical

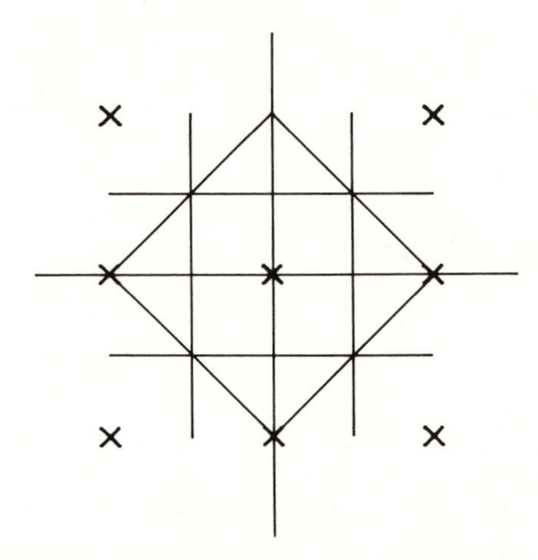

Fig. 2.1 First Brillouin zone for square lattice.

with that for an electron in the field of a single atom. The function (2.1.1) then reduces approximately to

$$\Sigma_j \exp(i\mathbf{k}\mathbf{R}_j).\phi(\mathbf{r} - \mathbf{R}_j). \qquad (2.1.3)$$

Here $\mathbf{R}_j$ is the position of the jth atom, and ϕ the atomic wave function. (We assume a *Bravais lattice,* i.e., a lattice in which the position of any atom can be reached from a given one by a symmetry operation.)

It is then easy to obtain the approximate form of the energy bands, and in one dimension this looks like figure 2.2. In particular the energy has a horizontal slope at both ends, so that the left-hand boundary fits smoothly on to the right-hand one. The same is true of the wave functions. This must be so because in one dimension the phase factor exp ika, which gives the relative contributions of successive terms to (2.1.3), approaches -1 as k approaches either $+\pi/a$ or $-\pi/a$. Similarly, there is continuity in two or three dimensions between one boundary of the first zone and the opposite boundary. For example, for the square lattice whose Brillouin zone is shown in figure 2.1, the factor in (2.1.3) equals $(-1)^j \exp ik_y Y_j$ both on the right- and left-hand edge of the zone.

I used this behavior in 1929 to explain the positive sign of the Hall effect in certain metals. In modern language the idea can be

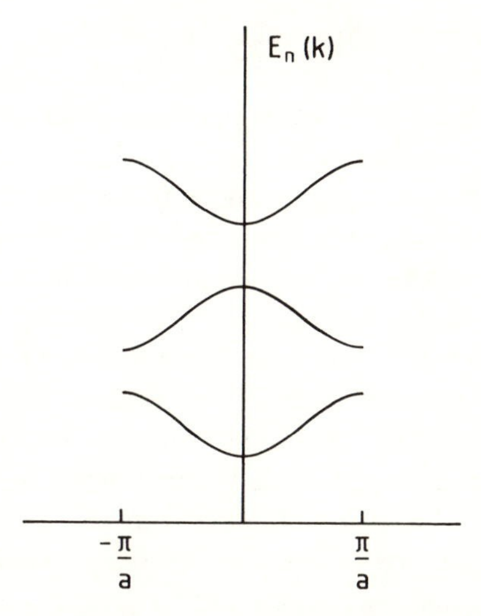

Fig. 2.2 Bloch energy curves for tight binding.

expressed by saying that if a band is nearly full, the state of the electrons can be described by specifying only the empty states, which will be at the top of the band. Since the energy is near the top, a decreasing parabolic function of the distance of **k** from the point of maximum energy, the energy change by vacating a state is a rising parabolic function, thus giving normal kinematics for the "holes". The vacancy also carries a positive charge (absence of a negative charge), and one can show that electric and magnetic fields affect it like a positive particle.

However, this explanation of the positive Hall effect could at that time be justified only in the limit of tightly bound electrons, since the behavior near the zone boundary had been investigated only for that limit. This was worrisome. Then it occurred to me to look at the case of nearly free electrons, which on the face of it seemed so very different from what I was looking for.

For completely free electrons it is most convenient to define the wave vector **k** to have all values, and the wave functions are then simply exp $i\mathbf{kr}$ with the energy proportional to k^2. However, it is also possible to write the wave function in the form of (2.1.1), since zero potential is also a periodic potential. For this we have to restrict **k** to the first zone and make u an exponential to correct for this. In one dimension this results in an energy curve like figure 2.3A.

Now assume there is a weak periodic potential. In general the effect of this can be obtained by perturbation theory, and it will not greatly change either the eigenfunctions or the energies. The magnitude of the perturbing effect is measured by the ratio between the matrix elements of the perturbing potential and the energy difference between the states connected by the matrix element. Since the periodicity is preserved, the matrix elements of the potential can mix only states with the same k, i.e., states lying vertically above each other in figure 2.3A. In most cases the energy differences are appreciable, but there are points where they vanish. This happens at $k = \pm\pi/a$, and also at $k = 0$, except at the lowest energy.

For those k values for which the difference vanishes exactly, we evidently have to use degenerate perturbation theory, which starts from a linear combination of the two degenerate functions. This suggested that where they are nearly degenerate we may still use a linear combination, but allow for the energy difference. We take the "function of zeroth approximation" as

$$a_1\psi_1 + a_2\psi_2, \tag{2.1.4}$$

where the ψ are exponentials. If we insert (2.1.4) in the Schrödinger equation and project the result on ψ_1 and ψ_2, we obtain

$$\begin{aligned} (E_1 - E)a_1 + V_{12}a_2 &= 0 \\ V_{21}a_1 + (E_2 - E)a_2 &= 0. \end{aligned} \tag{2.1.5}$$

We have omitted the diagonal elements of V, since these equal the space average of the potential, which can be made to vanish by a shift in the zero of energy.

The existence of a solution of the homogeneous equations (2.1.5) requires

$$E = \frac{E_1 + E_2}{2} \pm \sqrt{\left[\left(\frac{E_1 - E_2}{2}\right)^2 + V_{12}^2\right]}. \tag{2.1.6}$$

At the crossing point, where the two unperturbed energies coincide, we now find two resulting energies differing by $2V_{12}$, and near this point the energy changes are proportional to the square of the unperturbed energy difference, i.e., to the square of the distance of k from the crossover point. In other words, the energy curves now look like figure 2.3B. So we find that the curves look qualitatively like figure 2.2 in that they have a horizontal tangent at the edge. It is easy to see that the eigenfunctions fit smoothly from the one boundary to the opposite one.

This came as a great and pleasant surprise, since now the behavior was seen to be the same for very strongly and very weakly bound electrons, and therefore could be assumed to be general. I was satisfied with having the proof for one dimension, where the different "bands" arise from the free-electron states in the way made transparent by figures 2.3A and 2.3B.

In two and three dimensions the picture becomes more complicated. It was explored by Brillouin. For the two-dimensional square lattice the situation is illustrated in figure 2.4 (first drawn by Brillouin). The first zone is simple, as shown already in figure 2.1. The pieces making up the next few zones are numbered. Each has to be

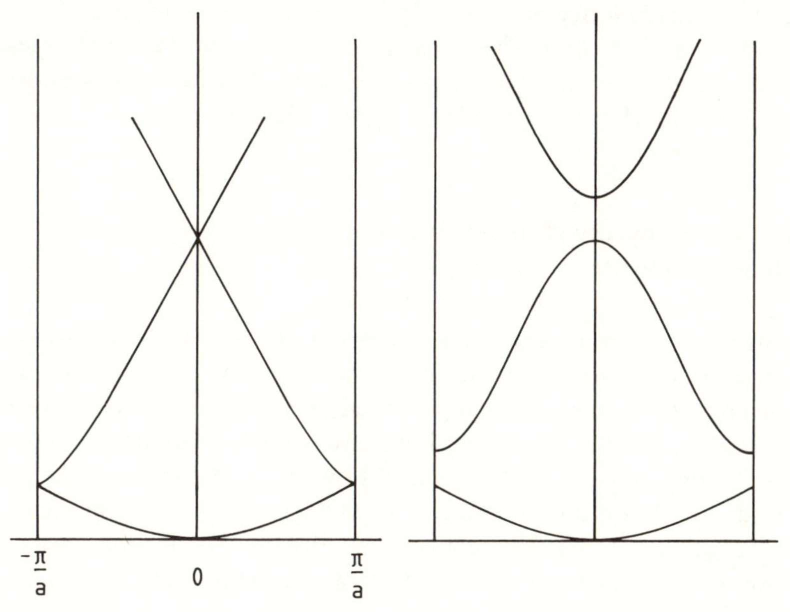

Fig. 2.3 Energy curves in one dimension. (A) *Left*: Free electrons. (B) *Right*: Weak periodic potential.

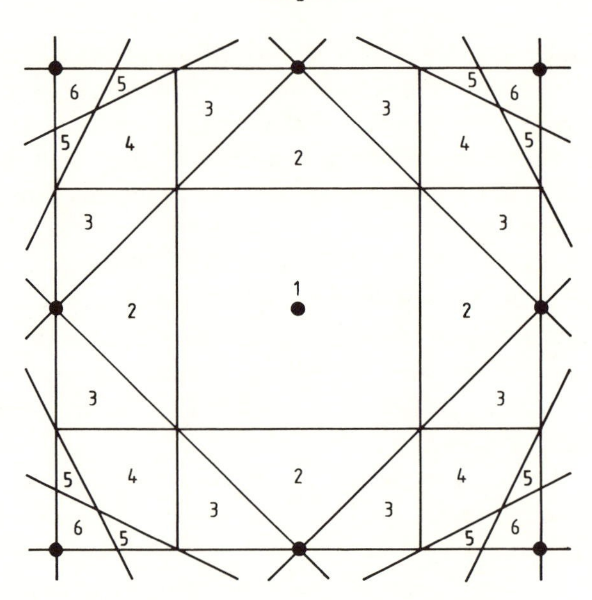

Fig. 2.4 Brillouin zones in two dimensions.

moved horizontally or vertically (or both) by a multiple of $2\pi/a$ to fit into the first zone. The joins are lines across which the free-electron energy has a discontinuous derivative, but this gets smoothed out just like the transition from figure 2.3A to 2.3B.

2.2 The Structure of Bismuth and the Jones Theory

In trying to understand the structure of crystals, it is natural to think of forces between the constituent atoms. In the first place, one thinks of an interaction potential between two atoms at a time, depending on their distance. Refinements are necessary to account for the dependence of the potential on the angle between different bonds involving the same atom, thus depending on the positions of at least three atoms. This concept of "directed bonds" has long been known to be essential for molecules and crystals, particularly those containing carbon atoms.

But from this point of view the structure of bismuth remained a complete mystery. This structure is close to a simple cubic lattice, but differs from it by every second atom being displaced a short distance along the space diagonal, and also the angle between the axes differs a little from 90 degrees. The structure is illustrated in figure 2.5, in which, however, the distortion has been exaggerated. Evidently no simple and plausible two- or three-body force law would make such a configuration stable. The solution to this paradox was provided by a surprisingly simple idea due to Harry Jones (1934a,b). The distorted lattice evidently has a lower symmetry than the original simple cubic one. In particular, there are now two sets of sites, which are not related by a symmetry operation, so that the unit cell now contains two atoms. In discussing this situation we shall ignore the small change in angle, which is a consequence of the other changes and does not affect the essence of the problem.

In this approximation we are then dealing with just a simple cubic lattice, with every second atom slightly displaced. The symmetry of the new lattice is that of a face-centered cubic lattice, with two atoms to the unit cell. The basic Brillouin zone therefore becomes that of the f.c.c. lattice, which is shown in figure 2.6. This is inscribed in the original cube of edge $2\pi/a$, which formed the Brillouin zone of the simple cubic lattice. The faces of the new zone that lie inside the cube are discontinuities produced by the symmetry-reducing potential, i.e., the change in the potential due to the displacement, in the manner discussed in the preceding section. The number of orbital states in the new zone is equal to the number of cells, and therefore half the number of atoms. If all lower bands

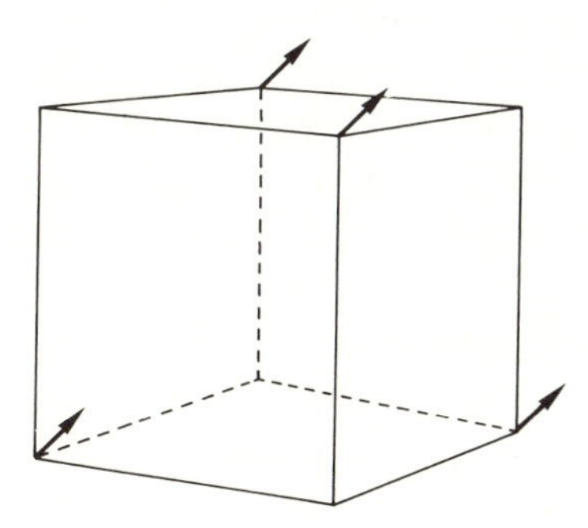

Fig. 2.5 Structure of bismuth, schematic.

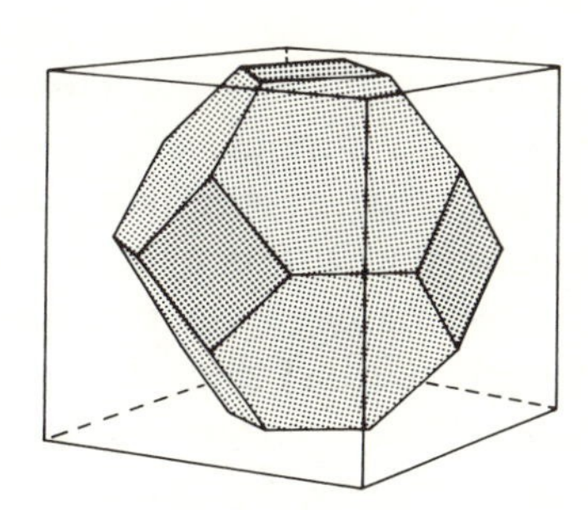

Fig. 2.6 Brillouin zone of f.c.c. lattice.

are filled, then, allowing for spin, the next subzone would have room for just one electron per atom.

Jones argued that the new surface of discontinuity illustrated in figure 2.6 was very near to a surface of constant energy in the undeformed lattice. This means that in bismuth, an element of odd atomic number (83) the Fermi surface would very nearly coincide with the new boundary, the valency electrons would practically fill the new sub-band.

The argument that Jones used to show that there was this near coincidence is based on the energy surfaces for nearly free electrons, and has for this reason been criticized. However, the consequences that follow agree with the facts in so many respects that the picture becomes very convincing.

In this picture most of the states just inside the new boundary are occupied; most of the states outside it are empty. The symmetry-breaking potential raises the energy of the states outside and lowers the ones inside (cf. fig. 2.3). The net result is a lowering of the total electron energy. In other words, the total electron energy tends to favor the displacement. The interaction between the ion cores will in general favor the more symmetric configuration. Whether the total energy of the distorted lattice is higher or lower than that of the original simple cubic lattice depends therefore on the relative magnitude of the two effects, but it is evidently possible that the distorted lattice is the stable configuration.

If the coincidence of the Fermi surface and the zone boundary was exact, one band would be completely filled and the next higher one completely empty, making the substance an insulator. In actual fact a few electrons will spill over into the next higher band, leaving a few holes in the lower one. Figure 2.3B also illustrates the fact that these extra electrons and holes are in regions of very great curvature of the energy surface, which means a very low effective mass. Since the electron diamagnetism is inversely proportional to the square of the mass (see *Surprises*, eq. [4.3.10]), this accounts for the exceptionally large diamagnetism of bismuth. The analysis of the de Haas–van Alphen effect (*Surprises*, sec. 4.4) in bismuth also shows a surprisingly small number of conduction electrons, with a very low effective mass, and great anisotropy. All these features are naturally accounted for by Jones's picture.

While the argument convincingly explains how such a strange structure can be stable, i.e., represents a local energy minimum, and has a lower energy than the simple cubic structure, it does not prove that it has a lower energy than other more conventional structures. Jones invokes a similar picture for the "Hume-Rothery" alloys, which include β brass. These also have structures of a lower symmetry than one would expect from simple chemical forces, and a suggestive point is that, comparing the concentrations for which such a structure occurs in different alloy systems, one finds that the concentrations are not the same, but that the ratio of the number of valence electrons to atoms is the same. This is of course required on the Jones picture because the electrons must be capable of filling a given sub-zone.

2.3 Peierls Transition

The fact that the total electron energy may favor a reduction of symmetry takes a particularly extreme form in one dimension, where both the edge of the Fermi distribution and the position of the discontinuity produced by the symmetry-reducing deformation consist of single points, so that there is no problem in making them coincide exactly.

Consider a linear chain of atoms, with a regular spacing a. For an odd number of electrons the valency band will be half full. The band energy may have the shape of figure 2.7A, which also shows the Fermi wave number $k_F = \pi/2a$ for one electron per atom in the band. Now let us displace every second atom by a small distance δ. This reduces the symmetry to that of a chain with spacing $2a$, and the potential acquires a Fourier component of wave number π/a, which in this case is equal to $2k_F$.

By the mechanism discussed in section 2.1 this results in a discontinuity of the energy curve, as in figure 2.7B. In this case all states raised by the change are empty, and all states lowered are occupied, so there is evidently a net reduction in electron energy.

The electron energies in the distorted chain are again given by (2.1.6), where now V is the matrix element of the change in the

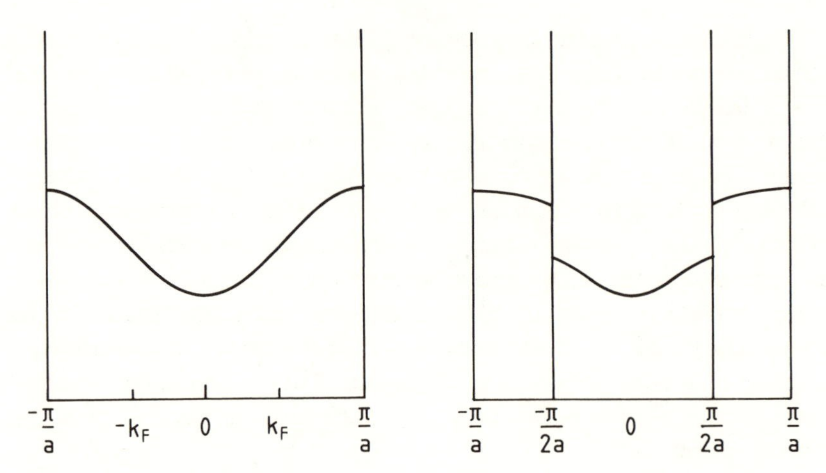

Fig. 2.7 Electron energy in linear chain. (A) *Left*: Uniform spacing. (B) *Right*: Every second atom displaced.

potential due to the displacement. There is no diagonal element. The reduction of the lower energy is, from (2.1.6),

$$\Delta(k) = E_1 - E =$$

$$\frac{-(E_2 - E_1)}{2} + \sqrt{\left[\frac{(E_2 - E_1)^2}{2} + V^2\right]}. \qquad (2.3.1)$$

To get the reduction of the total electron energy, we have to integrate this expression from $-\pi/a$ to $+\pi/a$, multiplied by $L/2\pi$, the number of electron states per unit k. The main contribution comes from the neighborhood of $\pm\pi/2a$, where we may regard V as constant, and $E_2 - E_1$ as linear in k. We write near $+\pi/2a$

$$E_2 - E_1 = \alpha(\pi/2a - k) \equiv \alpha\kappa. \qquad (2.3.2)$$

The approximations are valid only in the neighborhood of $\kappa = 0$. We must therefore restrict the integration to a maximum value of κ (minimum value of k), say κ_1, such that

$$V/\alpha \ll \kappa_1 \ll \pi/2a. \qquad (2.3.3)$$

The two regions near $k = \pm\pi/a$ give equal contributions. The integral is elementary and yields

$$\Delta E_{\text{total}} = \frac{V^2}{\alpha}\left[\left(\frac{V}{\alpha\kappa_1}\right)^2 + \frac{1}{2}\log\frac{\alpha\kappa_1}{V} - \frac{1}{4}\right]. \qquad (2.3.4)$$

The important feature of this result is that it behaves for small V as $-V^2 \log V$. For small displacements, V is proportional to the displacement δ. The behavior of the reduction in electronic energy for small displacement is therefore as

$$- \delta^2 \log \delta. \tag{2.3.5}$$

This is interesting because there may be other effects favoring the regular spacing, $\delta = 0$, such as the repulsion between the atomic cores, but these will have an energy varying as δ^2, and therefore the electron energy must dominate for small displacement. This suggests that the periodic chain must always be unstable.

This instability came to me as a complete surprise when I was tidying material for my book (Peierls 1955), and it took me a considerable time to convince myself that the argument was sound. It seemed of only academic significance, however, since there are no strictly one-dimensional systems in nature (and if there were, they would become disordered at any finite temperature; see *Surprises*, sec. 4.1). I therefore did not think it worth publishing the argument, beyond a brief remark in the book, which did not even mention the logarithmic behavior.

It must also be remembered that the argument relies on the adiabatic approximation, in which the atomic nuclei are assumed fixed. If their zero-point motion were taken into account, the answer might change, but this would be a difficult problem to deal with, since it involves a strongly nonlinear many-body problem.

In recent years distortions of the kind discussed above have been found in many crystals consisting of linear chains of molecules tied together rather weakly, which are therefore quasi one-dimensional. These results have been associated with my old result. While there is undoubtedly some analogy, there is the difference that in a three-dimensional crystal the Fermi surface is necessarily curved, and therefore cannot coincide precisely with the plane of discontinuity, though it may be very elongated in the directions at right angles to the chains. The logarithmic term in (2.3.3) is then absent, and with it the conclusion that the periodic arrangement *must* be unstable. It *may* be unstable, as was already clear from the Jones theory of section 2.2. The quasi one-dimensional structure facilitates the near-coincidence between the Fermi surface and the plane of discontinuity.

In this case the first surprise was in the mathematical result for the one-dimensional case; a second surprise was that this had some connection with reality.

2.4 Momentum and Pseudomomentum

Every physicist is familiar with momentum and with the circumstances in which it is conserved. A similar and also very important quantity, the *pseudomomentum* (sometimes called quasi-momentum, crystal momentum, or wave vector), is not nearly as well known. It was mentioned briefly in section 4.2 of *Surprises*, but there is more to be said.

We remember the definition of pseudomomentum: any conserved quantity in physics is associated with some symmetry of the physical laws, in particular of the Hamiltonian. If the physics is invariant under the displacement of the origin,

$$x,y,z \rightarrow x + a,y,z, \tag{2.4.1}$$

the law of conservation of momentum is valid. Pseudomomentum is associated with a differnt transformation:

$$f(x,y,z) \rightarrow f(x - a,y,z), \tag{2.4.2}$$

where f is any physical quantity defined at each point in the medium under consideration. This replacement differs from (2.4.1) by keeping the medium fixed, but moving its physical state. In vacuum, when there is no medium, the two transformations are identical, and then pseudomomentum and momentum coincide.

For the physics to be unaffected by the replacement (2.4.2), the medium must be homogeneous. For solids, in which there is an atomic structure, pseudomomentum is a macroscopic concept. The transformation (2.4.2) is applicable to perfect crystals, but only with the displacement a being a multiple of the lattice constant. As a result, pseudomomentum is then not conserved, but can change by discrete amounts, $\hbar$ times any reciprocal lattice vector of the crystal, as in Bragg reflections or in Umklapp processes.

We can write a formal expression for the pseudomomentum. Let $q_\alpha(r)$ be a set of variables specifying the state of the medium at a point $\mathbf{r}$, and let π_α be their canonically conjugate momenta. Then the ith component of pseudomomentum is

$$K_i = -\sum_\alpha \int d^3\mathbf{r}\pi_\alpha(\mathbf{r}) \frac{\partial q_\alpha}{\partial r_i}. \tag{2.4.3}$$

We assume that the q_α vanish or approach constant values at large distance, so that the integral converges.

We verify that the definition connects correctly with (2.4.2). For this purpose we form the Poisson bracket

$$\{K_i, f\} \equiv \sum_\alpha \int d\mathbf{r}\, d\mathbf{r}' \left[\frac{\partial K_i}{\partial q_\alpha} \frac{\partial f}{\partial \pi_\alpha} - \frac{\partial K_i}{\partial \pi_\alpha} \frac{\partial f}{\partial q_\alpha} \right]. \tag{2.4.4}$$

Classical Poisson brackets are perhaps less familiar today than their quantum equivalent, commutators. However, since the present discussion is essentially classical I prefer to use the language of Poisson brackets. The derivatives of $\mathbf{K}$ are functional derivatives, since K depends on the field of π's and q's.

From (2.4.3),

$$\frac{\partial K_i}{\partial \pi_j} = -\frac{\partial \pi_j}{\partial r_i}; \quad \frac{\partial K_i}{\partial q_j} = \frac{\partial \pi_j}{\partial r_i}. \tag{2.4.5}$$

(For the second relation one has to integrate the definition [2.4.3] by parts; the boundary term vanishes because we have assumed the q or π to vanish at large distance.) Inserting in the definition of the Poisson bracket, (2.4.4),

$$\{K_i, f\} = -\sum_j \left(\frac{\partial f}{\partial q_j} \frac{\partial q_j}{\partial r_i} + \frac{\partial f}{\partial \pi_j} \frac{\partial \pi_j}{\partial r_i} \right) = -\frac{\partial f}{\partial r_i} \tag{2.4.6}$$

This is the change in f brought about by an infinitesimal transformation (2.4.2). If f is left unchanged by this transformation, its Poisson bracket with $\mathbf{K}$ vanishes.

By assumption this applies to the Hamiltonian. The relation

$$\{K_i, H\} = 0 \tag{2.4.7}$$

therefore expresses both the fact that the Hamiltonian is unaffected by the displacement and that the time derivative of $\mathbf{K}$ is zero, so $\mathbf{K}$ is conserved.

In quantum theory $\mathbf{K}/\hbar$ is the wave vector, whose importance in many condensed-matter problems has been recognized for a long time; but the surprise is that in many applications in which both momentum and pseudomomentum are conserved, the latter turns out the simpler and more useful quantity, as we shall see in a few examples.

One example of this definition is the pseudomomentum of a fluid. Since the definition (2.4.3) is expressed in terms of canonical variables, we must use a canonical description of the fluid, which is possible in the so-called Lagrange variables. These describe the present position $\mathbf{r}$ of an element of fluid, labeling the element by a reference position $\mathbf{R}$, which is usually chosen as the position of this element at a time when the fluid was at rest at uniform density. It is also convenient to introduce the displacement,

$$u_i = r_i - R_i. \tag{2.4.8}$$

In these variables (2.4.3) becomes

$$K_i = -\rho_0 \int d^3\mathbf{R} \sum_j v_j \frac{\partial u_j}{\partial R_i}. \tag{2.4.9}$$

Here ρ_o is the (constant) density in the reference state and $\mathbf{v} = d\mathbf{r}/dt$ is the velocity. If the fluid fills all space, we require that the integral (2.4.9) converge, which is true if the velocity or the gradient of the displacement tends to zero at infinity.

It is easy to verify that (2.4.9) is indeed conserved, if there are no forces acting on the fluid, and the pressure on the boundary, if any, is constant. We write the rate of change of K_i as

$$\dot{K}_i = -\rho_0 \int d^3\mathbf{R} \sum_j \left(v_j \frac{\partial u_j}{\partial R_i} + \dot{v}_j \frac{\partial v_j}{\partial R_i} \right). \tag{2.4.10}$$

The equation of motion of the fluid is

$$\rho \dot{v}_j = -\frac{\partial p}{\partial x_j} + \rho F_j, \tag{2.4.11}$$

where ρ is the actual density, F is the force per unit mass, and the derivative is with respect to the actual space coordinates. The element of mass must be the same both in terms of the actual, as of the reference coordinates,

$$\rho_0 \, d^3\mathbf{R} = \rho d^3\mathbf{r}, \tag{2.4.12}$$

and (2.4.10) becomes

$$\dot{K}_i = \int d^3\mathbf{r} \frac{\partial p}{\partial r_i} - \rho_0 \int d^3\mathbf{R} \left[\frac{\partial P}{\partial R_i} + \frac{1}{2} \frac{\partial v^2}{\partial R_i} + \sum_j F_j \frac{\partial u_j}{\partial r_i} \right]. \tag{2.4.13}$$

Here P stands for the usual "pressure function,"

$$P = \int \frac{dp}{\rho}. \tag{2.4.14}$$

The integrals are to be taken over the whole volume in actual space and in the reference space, respectively. All but the force term are gradients that reduce to differences of boundary values. Hence in the absence of an external force and for constant pressure on the boundary the expression vanishes.

For a motion with small amplitude, the pseudomomentum is evidently of second order in the amplitude. In the case of a sound wave of small amplitude, one easily verifies that

$$K = E/s, \tag{2.4.15}$$

where E is the energy of the wave and s the velocity of sound. In quantum mechanics, where the energy of a phonon is $\hbar ks$, this becomes $\hbar k$, where k is the wave vector.

It can also be shown that, in the case of a rigid body immersed in the fluid, the sum of the pseudomomentum of the fluid and the momentum of the body is conserved, a result that is useful for many applications. This follows by noticing that in the case of an immersed body the integral in the first term of (2.4.13) excludes the volume of the body, so that boundary terms remain, which amount to minus the net force on the body. Hence the sum of the pseudomomentum of the fluid and the momentum of the body is conserved.

In particular, the thrust on an immersed body due to the reflection of a sound wave can be obtained immediately as the difference of the pseudomomentum of the wave before and after the reflection. One might expect that it should be obtained by considering the change of the momentum of the fluid. We shall see in the next section, however, that the momentum of a sound wave is a very complicated quantity. Moreover, it would not be correct to equate the thrust to the momentum difference between the incident and the scattered wave, because as a consequence of the nonlinearity of the equations of fluid dynamics, transient motions are generated in the act of reflection, and their momentum cannot be neglected. Their pseudomomentum is, however, negligible in most cases, since their wavelength is long, of the order of the extension of the sound pulse. Hence the pseudomomentum is not only the simpler quantity, but it is more useful.

The same applies in many other applications. We shall meet in setion 2.6 another example in which physical reasoning required the use of pseudomomentum.

Historical note: The expression (2.4.9) for the pseudomomentum of a fluid has been well known for some time. I expect that also the more general expression (2.4.3) can be found in the literature, but I do not know any specific reference. The argument that conservation of pseudomomentum is valid and useful even in the case of a body immersed in a fluid was given in a letter of May 1981 to me by Dr. M. E. McIntyre (Cambridge). I failed to acknowledge this in my 1983 Varenna lectures (Peierls 1984) because I had forgotten that it was mentioned there, and I apologize to Dr. McIntyre for this apparent plagiarism.

2.5 Momentum of a Sound Wave

Surprises, section 4.2, dealt with the momentum of a phonon, or of a sound wave, in a medium in which the equations of motion showed dispersion but were still linear. The discussion led to the conclusion that a sound signal that did not split in two in the course of its propagation necessarily had zero momentum. But since the discussion had made no allowance for nonlinearities, I pointed out that further surprises could not be ruled out. Indeed, a study of nonlinear effects leads to a new surprise.

To see this we consider a fluid in which the nonlinear effects and the dispersion are weak enough to be considered only as first-order corrections. We assume, accordingly, a density of potential energy:

$$V = V_0 + \frac{c^2}{\rho_0}\left\{\frac{1}{2}\left(\frac{\rho_0 - \rho}{\rho_0}\right)^2 + \frac{\beta}{3}\left(\frac{\rho_0 - \rho}{\rho_0}\right)^3 - \frac{1}{2}\alpha\left(\frac{\nabla\rho}{\rho_0}\right)^2\right\}. \tag{2.5.1}$$

Here c is the velocity of sound, α is a measure of the dispersion, i.e., the dependence of wave velocity on wavelength, and β is the nonlinearity. ρ_0 is the fluid density in the reference state used in the Lagrange description.

The ratio of this to the actual density is then equal to the Jacobian of the transformation from the Lagrange reference coordinates to the space coordinates, which is, up to second-order terms,

$$\frac{\rho_0}{\rho} = 1 + \operatorname{div}\mathbf{u} + \frac{1}{2}(\operatorname{div}\mathbf{u})^2 - \frac{1}{2}\sum_{l,m}\frac{\partial u_l}{\partial R_m}\frac{\partial u_m}{\partial R_l}. \tag{2.5.2}$$

Here $\mathbf{u}$ is the displacement, as in (2.4.8). The *div* is taken with respect to the Lagrange reference coordinates R_i.

This leads to the Lagrange equation of motion to the required order

$$\frac{\partial^2 u_i}{\partial t^2} = c^2\frac{\partial}{\partial R_i}\operatorname{div}\mathbf{u} + \alpha\frac{\partial}{\partial R_i}\nabla^2\operatorname{div}\mathbf{u} + \left(\frac{3}{2} + \beta\right)\frac{\partial}{\partial R_i}(\operatorname{div}\mathbf{u})^2$$
$$+\frac{1}{2}\left[\sum_{n,l}\frac{\partial u_l}{\partial R_n}\frac{\partial u_n}{\partial R_l} - \sum_n\frac{\partial}{\partial R_n}\left(\frac{\partial u_n}{\partial R_i}\,div\,\mathbf{u}\right)\right], \tag{2.5.3}$$

where the operators ∇^2 and div relate to the reference coordinates. Note that even for $\beta = 0$ there remains a nonlinear term in the equation, because of the kinematic nonlinearity contained in (2.5.2).

We now look for a solution representing a sound pulse. It contains a wave of wave vector k, frequency ω (therefore phase velocity $w = \omega/k$), group velocity g, and modulated by a factor f, which is slowly varying on the scale of wavelength. To allow for the possibility of the pulse carrying momentum, we also add a slowly varying nonoscillatory term h, which we require to travel with the pulse, not to run away from it.

In other words, we assume

$$\mathbf{u} = Re\{\mathbf{f}(\mathbf{R} - \mathbf{g}t)e^{ik(Z-wt)}\} + \mathbf{h}(\mathbf{R} - \mathbf{g}t). \tag{2.5.4}$$

The vectors **f**, **w**, **g** are assumed in the same, say the Z, direction. Inserting in (2.5.3) then leads immediately to

$$\begin{aligned} w &= c(1 - \tfrac{1}{2}\alpha k^2) \\ g &= c(1 - \alpha k^2), \end{aligned} \tag{2.5.5}$$

neglecting higher orders in u and k.

Now the condition for h becomes

$$g^2 \frac{\partial^2 h_i}{\partial Z^2} - c^2 \frac{\partial}{\partial R_i} \operatorname{div} \mathbf{h} = \frac{1}{2} c^2 \left\{ (1 + \beta) \frac{\partial}{\partial R_i} f_z^2 - \partial_{iz} \frac{\partial}{\partial Z} f_z^2 \right\}. \tag{2.5.6}$$

(Details of the derivation are given in Peierls 1983.)

This equation can be solved by a Fourier transform. Denoting the Fourier transform of **h** by $\phi(\mathbf{q})$ and that of f^2 by $\chi(\mathbf{q})$, we find

$$\phi_z = \frac{iq_z\chi}{g^2 q_z^2 (c^2 q^2 - g^2 q_z^2)} \left\{ \beta\, g^2\, q_z^2 + c^2\, (q^2 - q_z^2) \right\}. \tag{2.5.7}$$

The space integral of the z component of the momentum density is

$$-2\pi i \rho_0 q_z \phi_z g, \tag{2.5.8}$$

evaluated at $q = 0$. But, inserting from (2.5.7) we see that this expression has no limit at the origin; it depends on the ratio of q_z to q.

In coordinate space the momentum distribution, which is proportional to the Fourier transform of $q_z\phi_z$, behaves at large distances like

$$\frac{X^2 + Y^2 - \gamma^2 Z^2}{(X^2 + Y^2 + \gamma^2\, Z^2)^{5/2}}, \tag{2.5.9}$$

where

$$\gamma = (1 - g^2/c^2)^{1/2}. \tag{2.5.10}$$

This is similar to the field of a dipole (rather, a Lorentz-transformed dipole field) of which it is known that its integral over all space is ambiguous. If the integral is taken over a finite volume, and the size of the volume is made to grow indefinitely, the limit depends on the shape of the volume.

Consider in particular the choice of an extremely elongated volume, whose length in the Z direction is large compared to its width in the other directions. This corresponds to taking in (2.5.7), $q_z \ll q$. Then the total momentum becomes

$$\frac{E}{g(1 - \alpha k^2)}. \tag{2.5.11}$$

The second term in the bracket in the denominator is small, of second order, and therefore not significant in our approximation. Omitting it leaves $p = E/c$, which is the result that corresponds to a momentum $\hbar k$ for a phonon, and to that of a simple sinusoidal wave in the Euler description (see *Surprises*, sec. 4.2).

If, on the other hand, we choose a thin disk at right angles to the direction of motion, which corresponds to taking $q_z = q$ in the limit, we find the momentum

$$\frac{\beta\, g\, E}{3\, \alpha\, c\, k^2}. \tag{2.5.12}$$

This answer would be unreasonable physically because it diverges in the limit of infinite wavelength, and in the limit of no dispersion. This makes the alternative answer (2.5.11) more plausible, but I do not know of any argument to justify it. There is a slight analogy with an old problem in hydrodynamics, which will be discussed in section 7.1

The surprise is that the momentum of a sound pulse is a very complicated object and may not even have a well-defined value at all.

2.6 Momentum and Pseudomomentum of Light

We saw in section 2.5 that the momentum of a sound pulse is by no means a simple concept, while its pseudomomentum is a simple quantity and far more useful. The same turns out to be the case for light. There is no problem about the momentum of light in the vacuum (where it is also identical with pseudomomentum). It is well known that the momentum of a light wave is then

$$p_{\text{vac}} = \frac{E}{c}, \tag{2.6.1}$$

where E is the energy and c the light velocity.

However, the momentum in a medium of refractive index n (we shall neglect dispersion for simplicity) has caused a long controversy, starting with Minkowski's 1908 result,

$$p_{\text{M}} = \frac{nE}{c}, \tag{2.6.2a}$$

followed a year later by Abraham finding

$$p_{\text{A}} = \frac{E}{nc}. \tag{2.6.2b}$$

Before looking for the correct answer, we should be clear about the precise question to be answered. If light travels in a material medium, the atoms of the medium will be set in motion by the electromagnetic fields of the light. The momentum therefore consists of two contributions: that of the electromagnetic field and that of the moving atoms. We are concerned with the total amount. However, if we consider an infinitely extended plane wave in an infinitely extended medium, we can superimpose any uniform velocity, which will alter the total momentum. How do we dispose of that ambiguity?

One might suggest prescribing that the medium be at rest on the average, but there are different ways of taking the average, so this is not acceptable. There is, however, a unique answer if we consider a light pulse of limited extension. We may assume that ahead of the pulse the medium is at rest and undisturbed. As the front of the pulse passes, it sets the atoms of the medium in motion, and, as we shall see, when the pulse has passed, the atoms are again at rest to a good approximation (at least as far as motion in the direction of the light is concerned). The light pulse is thus accompanied by a mechanical momentum density.

The electromagnetic part of the momentum density, $g_{\text{e.m.}}$, is easy to determine. It is the time-space component of the relativistic stress tensor, which from general theorems is known to be symmetric. The component in question is therefore equal to the space-time component, i.e., the energy flux $S_{e.m.}$, apart from a factor $1/c^2$ which comes from the relativistic units. Thus the electromagnetic momentum density is

$$\mathbf{g}_{e.m.} = \mathbf{S}_{e.m.}/c^2 = \mathbf{E} \wedge \mathbf{H}/c^2 \qquad (2.6.3)$$

(the second equality is Poynting's theorem). By comparing this with the energy density, one easily verifies that this gives Abraham's relation (2.6.2b).

Abraham evidently believed that this applied to the whole momentum, and it is tempting to follow him by applying the symmetry argument to the whole momentum and the whole energy, which would give practically the same answer since the energy carried by the moving atoms is quite negligible.

However, the argument is relativistic, and for consistency the total energy density should include the density of rest energy, i.e., ρc^2, if ρ is the matter density. If the atoms are moving with an average velocity v, this adds a term $\rho v c^2$ to the energy flux, and by the symmetry relation a contribution ρv to the momentum density, a rather obvious result. (This involved my first surprise on this subject, after M. G. Burt and I had fallen into the trap of accepting Abraham's answer as the total momentum [Burt and Peierls 1973].)

As Abraham's answer includes only the electromagnetic momentum, and the mechanical momentum has to be added to it, one might guess that the total will turn out to be Minkowski's result (2.6.2a), but this is not the case. In fact Minkowski's result is the pseudomomentum, as was first shown by Blount (1971). One can verify this by applying the general result (2.4.3) to the electromagnetic field and the atoms of the medium. By the theorem stated at the end of section 2.4, the balance of the pseudomomentum can be used to predict the thrust on a body immersed in a fluid by a light wave. This means that one obtains correct predictions from the (incorrect) assumptions that Minkowski's result gives the momentum, and that the thrust balances the change of momentum of the light wave.

Returning to the momentum, it remains to determine the mechanical momentum, and this depends on the force exerted by the light pulse on the medium. The *ponderomotive force* in a nonmagnetic medium per unit volume is (see, e.g., Landau and Lifshitz 1960):

$$\mathbf{F} = \frac{1}{2}\epsilon_0 \nu \nabla E^2 + \frac{\epsilon - \epsilon_0}{2\epsilon_0 c^2}\frac{\partial}{\partial t}\mathbf{E} \wedge \mathbf{H}, \qquad (2.6.4)$$

where ν is an abbreviation for

$$\nu = \frac{\rho}{\epsilon_0}\frac{\partial \epsilon}{\partial \rho}. \qquad (2.6.5)$$

This is derived from a thermodynamic argument linking the mechanical force due to the electric field with the change of electric properties upon compression. Strictly speaking, (2.6.4) holds only for static fields, and one is tempted to think that the field of a light wave varies much too rapidly for a static approximation to be valid. It was therefore another surprise when Lal, Suen, and Young (1981) pointed out that the quadratic terms entering in the force have a time scale corresponding to the duration of the light pulse, which is slow enough in all practical cases.

The point is that, being quadratic in the field, the force produced by an approximately sinusoidal wave with frequency ω contains terms of frequency 2ω and terms of frequency zero, actually varying as the envelope of the pulse. The mechanical effect of the double-frequency terms is negligible, so that we are dealing with a slowly varying force. In the case of a progressive wave traveling in the z direction, we can replace $\partial/\partial z$ by $-n/c\, \partial/\partial t$, and E^2 is $(cn\epsilon_0)^{-1}\mathbf{E} \wedge \mathbf{H}$. So in this case

$$F_z = \left(n^2 - \frac{1}{2} - \frac{1}{2}\nu\right)\frac{\partial}{\partial t}(\mathbf{E} \wedge \mathbf{H})_z/c^2, \qquad (2.6.6)$$

so that, if $v_z = 0$ ahead of the light pulse, it is again zero when the pulse has passed, and there is a constant amount of momentum accompanying the pulse. This applies only to the longitudinal momentum; some transverse momentum is left after the pulse has passed, but its space integral is zero.

Also, the statement is restricted to a single progressive wave. On reflection or refraction, when there is also a reflected wave present, the interference between the waves will result in a force with non-vanishing time integral, leaving an acoustic disturbance after the light has passed. As this carries momentum, it is hard to draw conclusions from momentum conservation. The acoustic disturbance has, however, a negligible pseudomomentum, and therefore pseudomomentum conservation can be usefully applied.

The total momentum density is then, allowing for the electromagnetic momentum (2.6.3),

$$g = \left(n^2 - \frac{\nu}{2}\right) |\mathbf{E} \wedge \mathbf{H}|\hbar c^2. \tag{2.6.7}$$

One can get a rough idea of the magnitude by using the Clausius-Mossotti model, by which

$$\frac{\epsilon - \epsilon_0}{\epsilon + 2\epsilon_0} = \text{const. } \rho \tag{2.6.8}$$

so that

$$\nu = n^2 - 1 + \frac{1}{3}(n^2 - 1)^2 \tag{2.6.9}$$

and

$$p = \frac{1}{2}\left[(n^2 + 1) - \frac{1}{3}(n^2 - 1)^2\right]p_A. \tag{2.6.10}$$

It does not, however, seem possible in practice to verify this result by direct measurement.

This problem contained many surprises, but perhaps the greatest one was that many experiments appeared to verify Minkowski's prediction, although it had been proved that his expression for the momentum was not right. The explanation is that his expression give the pseudomomentum, and that conservation of pseudomomentum very generally can provide the right answer.

For example, in the experiments of R. V. Jones and collaborators (Jones and Richards 1954; Jones and Leslie 1978), in which the thrust on a mirror immersed in a liquid reflects radiation, conservation of pseudomomentum can be applied because of the theorem mentioned in section 2.4 concerning the case of an immersed body. The pseudomomentum of the light is given by the Minkowski expression (2.6.2), as was first pointed out by Blount (1971).

This is sometimes interpreted as showing that Minkowski's expression does give the momentum of the light, but momentum conservation cannot be simply applied, because in the reflection acoustic transients are created, which affect the momentum balance, though their pseudomomentum is negligible.

Historical note: There is an extensive and confusing literature on the subject, to which I contributed some right and some wrong answers. The paper with M. G. Burt, already mentioned (Burt and Peierls 1973), had a correct discusssion of the definition of the problem, but used the relativistic symmetry rule with the nonrelativistic expression for energy. A later paper (Peierls 1976) gave the right answer, except that it used the high-frequency value of the ponderomotive force. It agreed with the explanation of the recoil experiments by R. V. Jones et al., which had already been given by J. Gordon (1973). A further paper (Peierls 1977) gave correct expressions for the lateral force, but made wrong predictions about the case of oblique incidence by omitting the part played by the hydrostatic pressure in evening out the stress between the peaks and troughs of the wave. A lecture course in Varenna (Peierls 1984) contains a general overview, with correct answers, though with some copying errors in the derivation. A paper in the Gozzini Festschrift (Peierls 1987) discusses experiments with different geometries, but fails to find any case in which the answer would not be predicted correctly by pseudomomentum conservation.

2.7 Superconducting Sphere

In the 1930s we were puzzled by the problem of a superconducting sphere (or more generally an ellipsoid) exposed to a magnetic field. One knew that there was a critical field, H_c, such that for fields below this value, the induction B inside the body was zero, and if H exceeded H_c the substance became normal, i.e., $B = \mu_0 H$ (neglecting the very small susceptibility of a normal metal). All this, of course, related to what are now called Type I superconductors. Type II were then regarded as dominated by impurities.

Now, if a sphere is placed in a magnetic field below the critical value, the lines of force are pushed out from the body, and as a result the field on the surface is greater than the applied field. At the equator it is 50% higher. So the superconductivity should disappear when the applied field is $\frac{2}{3}H_c$. But if the body becomes normal, the field is then everywhere the external field, and not enough to destroy the superconductivity.

One's first guess is that part of the body might remain superconducting, perhaps an ellipsoid inscribed in the sphere. However, it would then be necessary for the field over its whole surface to be equal to H_c, and this is not compatible with the field equations. Other geometries can similarly be excluded. The phenomenological solution to this paradox can be seen (Peierls 1936) by plotting H as function of B in the superconductor. The information quoted so far would lead to figure 2.8. This leaves a gap in the curve from a to b. Clearly there must be an answer for intermediate values of B. The simplest possibility would be to join a and b by a straight horizontal line, as in figure 2.9, but evidently there are many other possibilities. They are restricted by the requirement that in the case of a long wire in a longitudinal field when H is given, no value between a and b should be thermodynamically stable, but this still allows a variety of shapes, such as figure 2.10.

This ambiguity can be settled by observations on a sphere or other ellipsoid. In that case there exists a well-known solution to the field equations, where

$$(1 - n)\mu_0 H + nB = \mu_0 H_{ext}. \qquad (2.7.1)$$

Here H and B are the field quantities inside the body; n is a coefficient; $4\pi n$ is known as the demagnetizing factor. For a sphere, n = 3/2.

This equation is represented by broken lines in figure 2.9, and the intersection with the $H(B)$ curve gives the physical solution. If the straight line from a to b is correct, the magnetic moment is

$$M = -H_{ext}|a^3, \text{ if } \frac{2}{3}H_c < H_{ext} < H_c\,. \qquad (2.7.2)$$

This is indeed confirmed by the experiments.

The result is more general than is evident from my presentation. Consider the thermodynamic potential in its dependence on the internal field. This is given by

$$Z = \int HdB. \qquad (2.7.3)$$

If we plot this against B, a well-known thermodynamic argument shows that any point on the curve where it is concave downwards is unstable and will break up into a two-phase region. This is because we can then draw a chord, which is below the curve, and since the two-phase system made up of two states given by ends of the chord is described by the chord, it will have a lower free energy.

Now suppose that the *B–H* diagram looks like figure 2.10. Then the thermodynamic potential looks like 2.11, and there exists a chord *O–D* below the curve, and a two-phase region will form, which reproduces figure 2.9.

In fact, Gorter and Casimir (1934) suspected that the intermediate state was such a two-phase region, though it did not seem possible to satisfy the boundary conditions between the layers. This is because there will be a transition region between the layers, which will affect the boundary conditions.

Later Landau (1937) showed by a microscopic theory that these layers did exist, and computed their thickness.

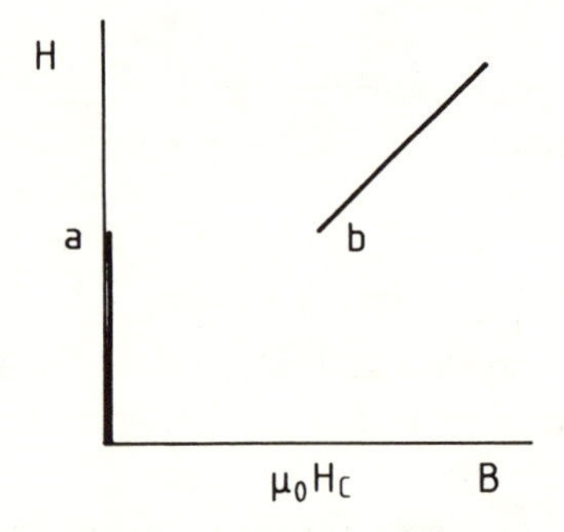

Fig. 2.8 Part of $H(B)$ diagram for superconducting sphere.

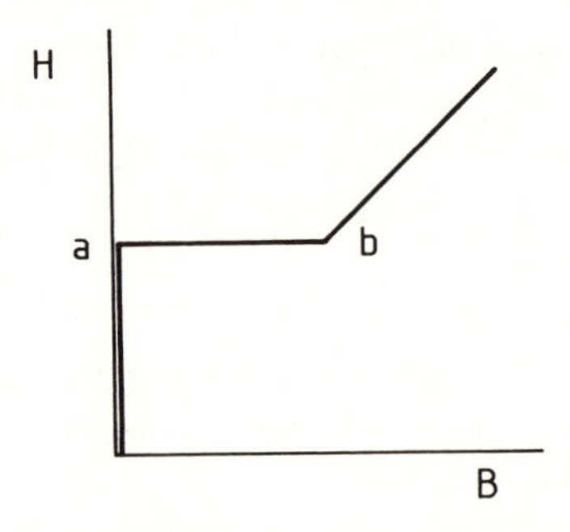

Fig. 2.9 Conjectured $H(B)$ diagram.

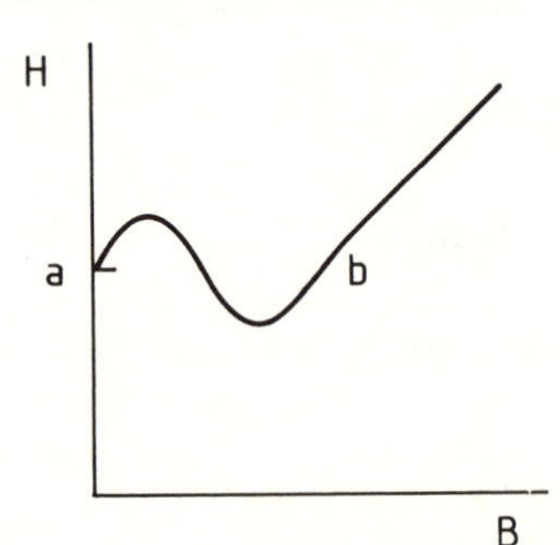

Fig. 2.10 Alternative $H(B)$ diagram.

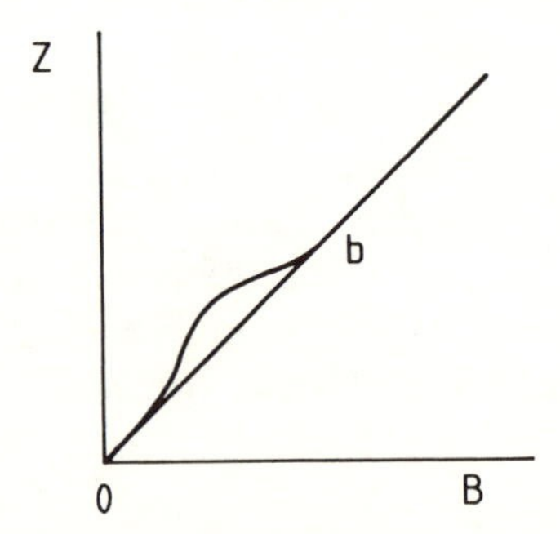

Fig. 2.11 Thermodynamic potential for the case of figure 2.10.

3

Statistical Mechanics

3.1 Bose-Einstein Condensation

It is well known that a perfect Bose gas, i.e., a gas of noninteracting molecules obeying Bose-Einstein statistics, shows condensation when the density exceeds a certain value, which depends on the temperature. This result is usually derived by using the grand canonical ensemble. This formalism is applicable in principle to a system in which the number of molecules is not fixed but is kept in equilibrium by exchange with a reservoir of molecules. One is usually interested in a quantity of gas enclosed in a vessel, so that the number of molecules is fixed. In normal circumstances the results obtained by means of the grand canonical ensemble are applicable nevertheless, because for a grand ensemble of macroscopic size the number of molecules fluctuates only by a negligible fraction, and therefore such an ensemble is practically indistinguishable from a canonical one. (The same argument is used to justify the use of a canonical ensemble, i.e., a system of given temperature, when in fact the system is isolated so that its energy is constant.)

Our first surprise here is that for a Bose system which shows condensation this step is no longer justified. To see this, consider the average occupation of each quantum state in the microcanonical ensemble:

$$n_k = (e^{\beta(\epsilon-\mu)} + 1)^{-1}. \qquad (3.1.1)$$

Here as usual $\beta = 1/k_B T$, ϵ_k is the energy of a molecule and μ the chemical potential. The latter should be chosen so as to make the expectation value of the total number of molecules take the right value:

$$N = \sum_{\mathbf{k}} n_{\mathbf{k}} \qquad (3.1.2)$$

For free particles,

$$\epsilon_{\mathbf{k}} = \frac{\hbar^2}{2m} k^2, \tag{3.1.3}$$

where the components of **k** take discrete values. For a vessel of macroscopic volume, the spectrum is very dense and we may try to replace the sum by an integral, with the element of integration

$$\left(\frac{L}{\pi}\right)^3 d^3\mathbf{k}, \tag{3.1.4}$$

allowing only positive values for the components of **k**. The behavior of this integral is well known. It is shown qualitatively in figure 3.1 as a function of μ. In particular, as μ approaches zero from negative values, N rises only to a finite limit N_0, whereas for $\mu \geqslant 0$ the integral diverges. Many elementary introductions merely say that the continuous distribution cannot accommodate more molecules than N_0, and the rest "fall out" in the bottom state. In a sense this is a correct statement, but in this form it sounds unconvincing.

As long as the integral approximation is valid, the result is either less than N_0 or infinity. For the approximation to break down, the difference between the occupation of adjacent levels must be appreciable. This happens first at the lowest levels whose energies are much less than $k_B T$, so that (3.1.1) becomes

$$n_k = k_B T/(\epsilon_k - \mu). \tag{3.1.5}$$

But we know that these states must contain between them the difference between the total number N and the number N_0 con-

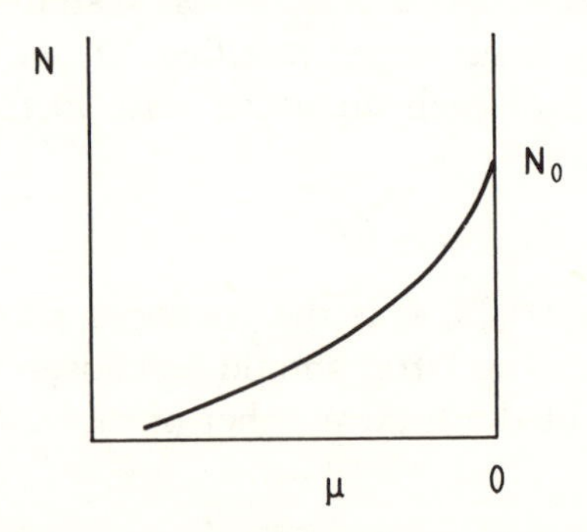

Fig. 3.1 Perfect Bose gas: particle number vs. chemical potential.

tained in the continuous region, which is a macroscopic number. So for at least some of these levels $(\epsilon - \mu)/k_B T$ must be extremely small, of the order of the inverse of a macroscopic number. But this can be true only for the lowest level, because ϵ_0 then is very close to μ, and for the next level the denominator is already of the order of the level spacing, which is inversely proportional to the square of the linear dimensions of the vessel, whereas $\epsilon_0 - \mu$ is inversely proportional to the volume. In other words, already n_1 is, for a macroscopic system, negligibly small, and the bulk of the excess of molecules really stays in the lowest state.

But how good is the grand canonical ensemble, which we have used? To judge this, we must estimate for this ensemble the fluctuations in the total number.

It is easy to show that the numbers in different states fluctuate independently, and for each

$$\langle \Delta n^2 \rangle = \langle n^2 \rangle - \langle n \rangle^2 = \langle n \rangle + \langle n \rangle^2, \tag{3.1.6}$$

so that the mean square fluctuation in the total number becomes

$$\begin{aligned} \langle \Delta N^2 \rangle &= \sum \langle \Delta n_k^2 \rangle = \sum \langle n_k \rangle + \sum \langle n_k \rangle^2 \\ &= N + \sum \langle n_k \rangle^2. \end{aligned} \tag{3.1.7}$$

In a macroscopic system, N is certainly small compared to N^2. In the absence of condensation, all the n_k are microscopic quantities, and the last term is therefore negligible compared to N^2, so the total number is practically constant. However, in the case of condensation, the ground state contains all the condensed molecules, which means that its contribution to the last term in (3.1.7) is $(N - N_0)^2$, and thus comparable to N^2. It seems therefore that the grand canonical ensemble is not equivalent to the canonical.

This objection is indeed justified if we insist on a pedantic application of the rule of the grand ensemble, i.e., if we want to determine the occupation number of *all* states, including the ground state, from knowledge of the chemical potential. But we have seen that the difference between the energy of the ground state and the chemical potential is minute, and we would never be able to know, or to hold constant, the chemical potential to a sufficient accuracy. Instead we just note that μ is very close to ϵ_0, which is enough to determine the occupation of all other states, and then use the knowledge of the total number to determine n_0. In other words, we apply the grand ensemble to the whole gas, *except* the condensate.

In this problem the main surprise is that we do best by using our naive intuition, and that a rigorous argument leads with much more trouble to the same answer.

3.2 The Ising Model

As a model of an order-disorder transition, Ising (1925) proposed a regular lattice of magnets, which we shall call *spins*, each capable of two orientations, say + or −, with an interaction energy of J for every pair of unlike neighbors. Clearly the lowest state is that where all spins are the same, when the total energy is zero. Ising also obtained the solution at any temperature of the one-dimensional system.

If we are interested in knowing the macroscopic limit, the answer is very easy to obtain. Let z_n be the partition function of a chain of length n, assuming n is large enough for the correlation between the two ends to be negligible. Then from every configuration of this chain we obtain a configuration of length $n + 1$ by adding another atom, which may either be like the last atom when it adds no further energy, or unlike it when it adds J, so

$$z_{n+1} = z_n (1 + x), \tag{3.2.1}$$

where

$$x = e^{-\beta J}; \ \beta = 1/k_B T. \tag{3.2.2}$$

Hence

$$z_n = c(1 + x)^n, \tag{3.2.3}$$

where c is a constant independent of n. It is easy to show that it has the value $2/1 + x$, but for our purposes this is of no importance.

The total energy is

$$E = -\frac{\partial}{\partial\beta} \log z = nxJ/1 + x. \tag{3.2.4}$$

In other words, the energy rises smoothly from 0 at zero temperature ($x = 0$) to $\frac{1}{2} nJ$ at infinite temperature ($x = 1$) when the spins are distributed at random. There is no sign of a phase change.

Ising believed that the same would apply in two or three dimensions, and others have produced sophisticated arguments for this belief. It seemed too difficult to locate the error in these arguments, but I was convinced it must be wrong and decided to prove the contrary, that there was indeed long-range order up to some finite temperature in two dimensions. The proof turned out to be surprisingly easy (Peierls 1935).

A configuration of the spins can be described by drawing boundaries separating the positive from the negative spins. The boundary determines the configuration, except that there are two possible configurations to a given set of boundaries, differing by exchanging the positive and negative spins with each other. An example of a spin system and its boundaries is given in figure 3.2.

The energy is given by

$$E = LJ, \tag{3.2.5}$$

where L is the total length of all boundaries, taking the lattice spacing as unit of length. The partition function can be expressed as

$$Z = 2 \Sigma G_L \exp(-\beta LJ), \tag{3.2.6}$$

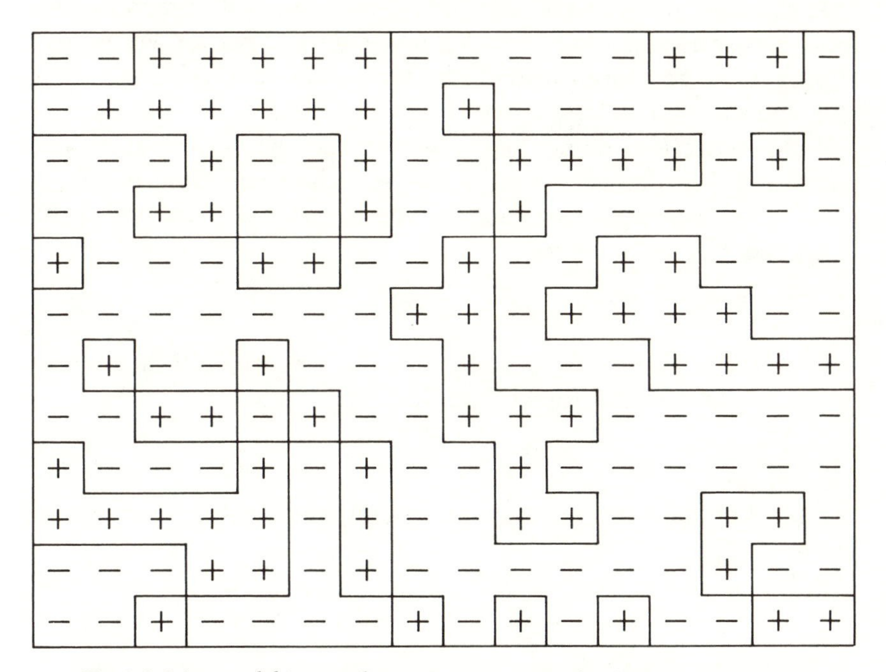

Fig. 3.2 Ising model in two dimensions: an example of a spin distribution.

where G_L is the number of ways in which we can draw boundaries of total length L. In drawing these boundaries we must observe the following rules:

No boundary may pass the same segment more than once.
No boundary may have a segment in common with another.
Each boundary must either be closed, or start and end on one of the edges of the array.
If two boundaries intersect, we can draw the line for each using either branch beyond the intersection. These alternatives must be counted only once.

These conditions make the evaluation of G_L a major problem. However, we can obtain an overestimate of the number of boundaries by ignoring all these conditions. This clearly overestimates the weight of configurations of high energy, and hence high disorder.

If the restrictions did not apply, we could draw boundaries starting from any point in the array and at each step would have four directions to choose from. (One could refine the estimate, because after the first step we have only three directions to choose from, and other restrictions could be taken into account. However, we are not aiming at an accurate estimate, for which the method is not suitable, but only for an inequality.)

If the restrictions do not apply, we can draw boundaries from any point, and the number of ways is just 4^s, if the boundary is of length s. Hence the partition function for the boundaries from one point would be

$$Z' = \Sigma(4x)^s. \tag{3.2.7}$$

If $4x > 1$, the sum diverges, so our estimates prove only the useless result that $Z < \infty$. But for $4x < 1$, we have

$$Z' = 1/1 - 4x, \tag{3.2.8}$$

and for the total expected length of boundaries starting from a given point,

$$\langle s \rangle = x \frac{\partial}{\partial x} \log Z' = 4x/1 - 4x. \tag{3.2.9}$$

How many spins can be enclosed by these boundaries, so as to make them different from the rest? Here we distinguish open and closed boundaries. The closed ones can be drawn from any point in the array. Each can enclose at most $(s/4)^2$ spin, so the total contribution of "wrong" spins is less than

$$N(x/1 - 4x)^2 \tag{3.2.10}$$

if N is the total number of spins in the array. The open boundaries can start only from points on the edge, i.e., from $4\sqrt{N}$ points. The area they can cut off is at most $(s/2)^2$, so their contribution is

$$4\sqrt{N}\,(2x/1 - 4x)^2, \tag{3.2.11}$$

which for macroscopic N is negligible compared with the closed boundary contribution (3.2.10).

It follows that, when $x/1 - 4x$ is less than $1/\sqrt{2}$, i.e., for $x < 0.18$, the number of wrong spins is less than half the total, so there is a nonvanishing magnetization. The surprise is that such a simple and crude argument can give a rigorous qualitative statement. The argument was later made unnecessary when Onsager produced an exact solution to the model, but it can be used in a variety of similar cases, when no exact solution is available.

4

Transport Problems

4.1 The Concept of Relaxation Time

In transport problems, and in the related problems of time-dependent nonequilibrium situations, one is often concerned with the approach to statistical equilibrium of a distribution differing only slightly from equilibrium. For example, we might consider the distribution of electrons in a metal (assumed noninteracting with each other) described by the number of electrons in a certain element of wave vector space,

$$f(\mathbf{k})d^3\mathbf{k}.$$

We assume the distribution slightly disturbed from equilibrium, e.g., by the presence of a weak current,

$$f(\mathbf{k}) = f_0(\mathbf{k}) + g(\mathbf{k},t), \tag{4.1.1}$$

where f_0 is the equilibrium (Fermi or Boltzmann) distribution, and g is treated as infinitesimal. The time dependence of g will show an asymptotic decrease to zero, and look, broadly speaking, like an exponential. It is therefore tempting to assume, for simplicity, that it is a simple exponential,

$$g(\mathbf{k},t) = g(\mathbf{k},0).e^{-t/\tau}, \tag{4.1.2}$$

independently of the form of g. The parameter τ is called the *relaxation time*. For a gas of particles whose equilibrium is restored by collisions, it is also often called the *collision time*. (In the older literature on kinetic theory one finds more often the *mean free path*, the average distance traveled between collisions, which in the then popular model of hard spheres was independent of velocity.) The appropriate differential equation for the time dependence of g is then

$$\dot{g} = -g/\tau. \tag{4.1.3}$$

To get a feeling for what one is neglecting in this approximation, we may consider the example of electrons that can make only elastic collisions with fixed centers, so that the collisions do not alter their energy. This is actually a good approximation even for the interaction with phonons at high temperatures, where the energy transfer to or from the phonons in one collision is small compared to $k_B T$.

The correct equation for g is then

$$\dot{g}(\mathbf{k}) = \int d\sigma' w(\mathbf{k},\mathbf{k}')[g(\mathbf{k}') - g(\mathbf{k})]. \tag{4.1.4}$$

Here the integration goes over a surface of constant energy in $\mathbf{k}$ space, and $d\sigma$ is the element of area on this surface; $w(\mathbf{k},\mathbf{k}')$ is proportional to the probability per unit time of scattering from state $\mathbf{k}$ to $\mathbf{k}'$ or vice versa. The *Boltzmann equation* (4.1.4) is in general a complicated integral equation, but it is easily solved in the case of isotropy, i.e., when the energy is a function of the magnitude of $\mathbf{k}$ only, and w depends, for a given energy, only on the angle between $\mathbf{k}$ and $\mathbf{k}'$. One can then expand g in spherical harmonics:

$$g = \sum_{l,m} g_{lm} Y_{lm}(\Theta,\phi), \tag{4.1.5}$$

and it is easy to show that from (4.1.4)

$$\dot{g}_{lm} = (w_l - w_0)g_{lm} = -g_{lm}/\tau_l, \tag{4.1.6}$$

where

$$w_l = 2\pi \int d\cos\Theta\, w(\Theta) P_l(\cos\Theta). \tag{4.1.7}$$

In this case we have a different relaxation time for different angular distributions of the deviation from equilibrium. Of particular interest is τ_1, which determines the transport properties, because in the presence of a current, g is proportional to $\cos\Theta$. This *transport collision time* is then given by

$$1/\tau_1 = 2\pi \int d\cos\Theta\, w(\Theta)(1 - \cos\Theta), \tag{4.1.8}$$

expressing the fact that small-angle collisions are unimportant for transport problems.

A specially simple case is *hard-sphere scattering*, for which w is independent of the scattering angle. Then all w_l vanish, except w_0, so that all relaxation times become equal, except τ_0, which is infinite. (The total number of electrons of a given energy is not affected by the collisions.) In that case we are dealing with a single relaxation time, but we must modify the equation (4.1.3) for g to read

$$\dot{g} = -\frac{1}{\tau}[g - \langle g\rangle], \tag{4.1.9}$$

where the brackets $\langle\ldots\rangle$ indicate an average over all states of a given energy. Even the more general form (4.1.6) is restricted to the isotropic case. Otherwise the exponentially decaying solutions of (4.1.4) generate an eigenvalue problem and in general the determination of the different relaxation times and the corresponding distributions may not be easy. If we allow energy transfer in the collisions, the problem is still more complicated.

One therefore often uses the simple equations (4.1.3) or (4.1.9) for orientation. However, one must guard against situations in which several times, perhaps of different orders of magnitude, play a part, depending on the nature of the deviation from equilibrium. A situation in which disregarding the variation of relaxation time and other parameters can give completely misleading results will be discussed in the next section.

In many problems in which the use of a single relaxation time would not do, one sometimes tries to get reasonable results by allowing τ to vary with the state of motion of the electron (or with the phonon mode, if one is concerned with phonons). One would then write in place of (4.1.4):

$$\dot{g}(\mathbf{k}) = -\frac{1}{\tau(\mathbf{k})}[g(\mathbf{k}) - \langle g\rangle]. \tag{4.1.10}$$

Here $\tau(\mathbf{k})$ would represent the mean life of a distribution in which only the occupation of the state $\mathbf{k}$ is disturbed from equilibrium.

But this approach may also be inadequate. An example is the problem of thermal conduction by phonons in the absence of Umklapp processes (*Surprises*, sec. 5.2). Here it is essential that the change in the occupation of one mode depends on the occupation of others. If only one mode was disturbed, it would return to equilibrium with a perfectly reasonable relaxation time; yet if the whole distribution is shifted in wave vector space, it will never return to equilibrium. This is the consequence of a conservation law, and we can deal with it in the same manner we dealt with the conservation of energy in writing (4.1.5). If the conservation law is only approximate, we do get involved with a genuine integral equation and might just as well start from the full Boltzmann equation without mention of a relaxation time.

It is clear that the indiscriminate use of the relaxation time concept can lead to many surprises. We can avoid them by a little reflection about the question we are asking and about the mechanism of relaxation.

One positive surprise was the fact that the Wiedemann-Franz law, which had been derived by using a single relaxation time, was in excellent agreement with experiment, at least at high temperatures, and at all temperatures for impure specimens. This law relates the electric conductivity σ and the thermal conductivity κ:

$$\frac{\kappa}{\sigma} = \frac{\pi^2 k_B^2 T}{3e^2}. \tag{4.1.11}$$

This is very easily derived on the assumption of a single relaxation time, and assuming only collisions without energy transfer. (Interactions with phonons involve energy transfers at most of the order of $k_B\Theta$, where Θ is the Debye temperature, and this is small compared to k_BT, and hence negligible, if $T \gg \Theta$.) The agreement was therefore at first taken as a justification of the relaxation-time approach. However, it can be shown that the law is much more general. The point is that the influences that cause the deviations from equilibrium—the external electric field and the temperature gradient—are both limited to the boundary region of the Fermi distribution, since the states with higher energy are empty and those with lower energy are filled, so they do not change. Over the Fermi surface both terms vary in the same way, so whatever the relaxation mechanism, the response of the distribution to these two disturbances is the same.

In an isotropic medium, where we are concerned with relaxation times τ_1, it is easy to see that both electric and thermal conductivity are governed by the same transport collision time τ_1. But we can derive the law more generally (see, for example, Peierls 1975).

So the surprise about the wide validity of the Wiedemann-Franz law should not encourage us to believe in a naive relaxation-time approach such as (4.1.3).

4.2 Magnetoresistance

If a metallic conductor is placed in a magnetic field, its electric resistance increases. We shall confine our discussion to the case of the magnetic field being perpendicular to the electric current. This phenomenon has received attention since the early days of the studies of metallic conduction, since it seemed likely that it would provide information about the mechanism of conduction. Interest was heightened when Kapitza showed that in most metals the resistance, after an initial rise with the square of the magnetic field, became linear in the field intensity.

When the modern electron theory of metals started developing in the 1920s, it was natural to attempt an explanation of this phenomenon. The first to try this was Sommerfeld, who used a simple relaxation-time approach. To his surprise the magnetoresistance came out negligible. This was later understood to be due to the fact that in his model all electrons had the same mobility, i.e., the same value of $e\tau/m$. Therefore, they acquire the same mean velocity in the electric field, and the magnetic field exerts the same deflecting force. Since the electrons are confined to the wire, a transverse electric field, the Hall field, is formed to stop the transverse current by canceling the magnetic deflecting force on the average. But since this force is the same for all electrons, the magnetic force is canceled for each electron, and their motion is independent of the magnetic field. In this model the magnetoresistance vanishes completely.

The fact that Sommerfeld found a small but non-zero contribution was due to the fact that he assumed a constant mean free path, and therefore a relaxation time inversely proportional to the velocity. The electrons contributing to the conduction are those in the boundary region of the Fermi distribution, which have nearly the same velocity, but there is a range proportional to k_BT/E_F, giving rise to magnetoresistance proportional to $(k_BT/E_F)^2$, a very small quantity in normal metals.

In real metals there are several causes that vary the mobility. In some metals the conduction electrons are spread over several bands, and the properties in different bands will differ. A specially transparent example is one where there are electrons in one band and holes in another, with equal numbers, masses, and relaxation times.

Then the transverse currents will cancel without any Hall field, but both electrons and holes will experience the effect of the magnetic field. (The model assumes that electron-hole pairs can be created and annihilated at the boundaries.)

In many metals there is only one conduction band involved, and then another cause of variation is the anisotropy of the electron motion because of the lattice structure. Electrons moving in different directions may have a different effective mass. This was for a time believed to be the main cause of magnetoresistance, but it was then surprising that the alkalis, which have substantial magnetoresistance, have Fermi surfaces that are extremely accurate spheres, so there is not much room for anisotropy.

There is, however, a further cause that contributes to the magnetoresistance in all metals, and that is the anisotropy of the phonons. For a small number of electrons in a band when the effective-mass approximation is adequate, the electron energy is a quadratic function of the wave vector. But for cubic crystals, a quadratic function respecting the crystal symmetry must be isotropic. The phonon spectrum is determined by the elastic properties. The elastic constants represent a tensor of the fourth rank, which even in cubic symmetry does not have to be isotropic. Hence the sound velocity varies with direction even in a cubic crystal. In other words, the phonons are much more sensitive to the crystal anisotropy than the electrons.

So the surprise about the order of magnitude of the effect is resolved. That leaves the surprise of the linear law, which was explained after many years by the work of Ilya Lifshitz and collaborators (Lifshitz et al. 1957).

The basis of their theory is that in the periodic field of the crystal one can introduce a gauge-invariant vector,

$$\kappa = \mathbf{k} - (e/\hbar)\mathbf{A}, \tag{4.2.1}$$

where **A** is the vector potential of the magnetic field. These new momentumlike variables do not commute, but

$$[\kappa_x, \kappa_y] = (ie/\hbar)B_z, \text{ etc.} \tag{4.2.2}$$

The energy is then the same function of κ as without the magnetic field. If the field is not too strong, and the temperature not too low (roughly $\mu B \ll k_B T$), one can use a semiclassical treatment. The point representing the electron in the space then moves in a plane at right angles to the magnetic field and at constant energy.

These contours can be of two kinds: they can be closed, being confined in one cell of the reciprocal lattice, or open, running from one face of the cell to the opposite and by continuation indefinitely through the reciprocal lattice. The nature of the orbits depends on the band structure and on the direction of the magnetic field. In general, open orbits exist only for very special directions of the field relative to the lattice. If the field is strong enough for the Larmor period to be smaller than the collision time, the magnetoresistance is proportional to B^2 for open orbits and saturates for closed ones. The quadratic behavior persists for small deviations from the direction giving open orbits, but the solid angle in which this occurs is proportional to $1/B$. In a polycrystalline sample all directions get averaged out, and therefore the overall resistance goes like B^2 times $1/B$, i.e., linearly. The behavior predicted by the Lifshitz group for a single crystal was in fact confirmed by experiment.

4.3 Electromigration

It has been known for a long time that hydrogen is easily dissolved in certain metals, such as palladium. It has also been known that an electric field applied to the metal will drive the hydrogen in the field direction, i.e., in the direction in which it would drive a positive charge. This was interpreted as showing that the hydrogen was in the form of protons and not of neutral atoms, which today we would take for granted. In the following we shall consider an ion of charge Ze ($Z = 1$ for hydrogen).

The motion of the ion evidently depends on the force it experiences and on the friction or other resistance to its motion. The latter may involve complicated problems, but the question of the driving force would seem very simple. However, the field acting on the ion contains, beside the external field, also that due to the conduction electrons, and a little reflection leads to the surprising result that the question is far from simple, and that even the sign of the force is not a priori obvious.

First of all, one has to worry about the possibility of screening. The ion will be surrounded by an increased electron density, so that the volume integral of the extra charge density equals $-Ze$ and therefore compensates the positive charge of the ion. If we were dealing with a hydrogen atom, in which an electron was bound to the proton, an electric field would evidently exert no force on the atom. Will the screening of the charge by the additional electron density cause a similar cancelation?

The question can be expressed in a different form: if the ion is moving with velocity u, and there is a force F acting on it, the force is doing work at the rate $F.u$. On the other hand, the rate of work done by the electric field is the volume integral of $E.j$ (E = field intensity, j = current density). The total current carried by the ion is $J = Zeu$, and if the extra electrons do not contribute to the current, energy conservation would say that $F = ZeE$.

The charge cloud must follow the ion, so there must be some current distribution, but this can take different forms. One extreme is the case in which the cloud is always made up of the same electrons, in which case the total electron current is $-Zeu$, and there is no resultant force; in the opposite extreme, the increase of the electron density ahead of the ion is made up from the decreasing density behind it, as sketched in figure 4.1. In that case, the total current due to the electrons vanishes and the force on the ion is ZeE. The second alternative is more plausible, because the electrons in a metal are also screened, and if such screening could cancel out the external force there could never be any conductivity.

Then, there is a further complication: the electric field gives rise to a current in the metal. The carriers of this current are scattered by the immersed ion, and, as this tends to reduce their forward pseudomomentum, it causes a reaction on the ion. This will be in

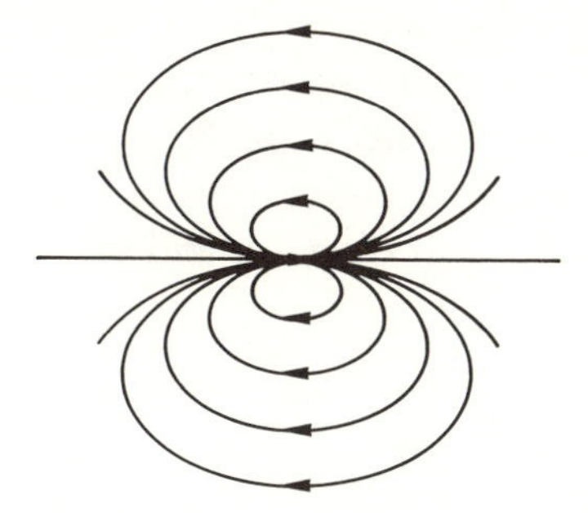

Fig. 4.1 Possible flow pattern near moving ion.

the direction of motion of the carriers, so if we are dealing with electrons carrying the current, the result will be a force opposite to the field direction, whereas in the case of hole conduction it will act in the same sense as the "direct" force. This force is often called the "electron wind" effect. There is no obvious way of estimating its magnitude compared to the direct force. It is therefore possible that in the case of electron conduction the hydrogen might move in a direction opposite to that of the field. (The use of pseudomomentum in this argument might appear an unnecessary sophistication; naively one might expect momentum conservation to give the answer. However, the electrons exchange momentum with the crystal lattice, so that momentum conservation contains an unknown amount. In the case of hole conduction this is particularly apparent; pseudomomentum conservation is simple because the total pseudomomentum of a filled band is zero.)

It will be clear from these remarks that the question is not trivial, and that it would be desirable to have a complete solution to the problem of the force. As far as I know there is no exact solution in the literature. There are many papers using simplified models, or dubious approximations. I shall here discuss the results of Das and Peierls (1973, 1975), based on a semiclassical treatment and the assumption of a constant collision time, and those of Sorbello and Dasgupta (1977), whose treatment is quantum mechanical but restricted to terms of the first order in the ion charge.

Das and Peierls start from a description in terms of classical dynamics, though with allowance for the Pauli principle, in the spirit of the Thomas-Fermi approximation. Moreover, they assume that the electrons are free and at each point in space have a probability $1/\tau$ per unit time of being scattered in a random direction. This is of course the constant-collision time, hard-sphere scattering approximation whose limitations were discussed in section 4.1.

In the frame of reference moving with the ion, the Boltzmann equation to first order in the external field E and the ion velocity u then takes the form:

$$\frac{\partial g}{\partial t} = -\frac{1}{m}\mathbf{p}\cdot\frac{\partial g}{\partial \mathbf{r}} - \frac{\partial \phi_0}{\partial \mathbf{r}}\cdot\frac{\partial g}{\partial \mathbf{p}} + \frac{1}{\tau}\left[\langle g\rangle - g\right] - \left(\frac{1}{m}\frac{\partial \phi_1}{\partial \mathbf{r}} - \frac{\mathbf{u}}{\tau}\right)\cdot\mathbf{p}\,\frac{\partial f_0}{\partial \epsilon}, \tag{4.3.1}$$

where f_0 is the undisturbed electron distribution in phase space, i.e., for $E = u = 0$; g is the first-order correction; and $\langle g \rangle$ its average over angles. ϕ_0 is the unperturbed electric potential and ϕ_1 its first-order correction, including the applied field. ϵ is the unperturbed electron energy. The first term on the right-hand side is the effect of the motion of the electrons, the second that of the acceleration by the field. The next term is the effect of collisions. The last term contains the deviation from the normal ion field (due to its motion and the applied field, and the deviation from isotropy of the scattering in the co-moving frame of reference).

ϕ_1 is given by the Poisson equation,

$$\epsilon_0 \nabla^2 \phi_1 = -e \int d^3\mathbf{p} \langle g \rangle = -e\rho_1, \qquad (4.3.2)$$

with the asymptotic behavior

$$\phi_1 \sim -Ez, \text{ as } r \to \infty. \qquad (4.3.3)$$

The unknown function g depends on two vectors, $\mathbf{r}$ and $\mathbf{p}$, i.e., on six variables. When we exploit the symmetry of the problem and take the electron gas as degenerate, the problem reduces to a function of two variables, but even in this form no solution has been worked out. There is, however, a useful identity: the solution for g remains unchanged if the velocity u is replaced by $u + v$, with arbitrary v, and the field by $E + (mv/e\tau)$, with ϕ_1 becoming $\phi_1 + (mv/e\tau)z$. Physically this means that in this model the effect of the field on the undisturbed metal is just to shift the electron velocity distribution in the z direction, and the solution can depend only on the velocity relative to that of the ion.

Since we are working only to first order in E and u, the force must be of the form

$$F = ZeE - CE - Du, \qquad (4.3.4)$$

with constant coefficients C and D. The identity requires a relation between C and D, so that

$$F = ZeE - D(u - u_E), \qquad (4.3.5)$$

where

$$u_E = eE\tau/m \qquad (4.3.6)$$

is the velocity of the conduction electrons.

The second term in (4.3.5) is evidently of the nature of the wind effect, so that in this model this is the only correction to the "direct" force, and the force is not screened off.

One can derive another theorem for this model by looking at the momentum balance. (Since the model assumes free electrons, there is no difference between momentum and pseudomomentum.) In the stationary state the momentum of the electron gas is constant, and therefore the force on the ion must balance the total force of the external field on all electrons and their loss of momentum by collisions. Since these quantities involve integrals over all volume, they must be evaluated carefully, using their asymptotic behavior. Das and Peierls find the result

$$F = -Nej_0\,\Delta\rho, \tag{4.3.7}$$

where j_0 is the current density in the absence of the ion, N the density of conduction electrons, and $\Delta\rho$ the change in resistivity due to the presence of one ion per unit volume.

If $\Delta\rho$ meant the increase of resistance due to scattering of electrons by the ion, then (4.3.7) would just be the expression for the "wind effect," and we would have the paradoxical result that this is the only force, and there was no "direct" term.

The solution to this paradox lies in an effect whose relevance was pointed out by Landauer (1975), the *carrier density modulation*. In the neighborhood of the positive ion there is an increased electron density (which is responsible for the screening), and as the model implies a local proportionality of the conductivity to the electron density, this results in an increase of the average conductivity, hence a negative contribution to $\Delta\rho$. If this is allowed for in (4.3.7) it causes a positive term in the force. For a weak ionic charge this can be shown to be exactly equal to the "direct" term, so there is no contradiction.

Das and Peierls give a solution for the case of a weak charge, retaining only the first-order terms in Z. The result gives a logarithmically divergent force, which is not surprising because the model uses classical dynamics, for which the average momentum transfer in the scattering from a point charge is divergent. However, one knows how this singularity is cut off by quantum effects and can therefore estimate the correct magnitude of the first-order term. This shows that even for a proton, with $Z = 1$, the first-order correction to F is greater than the direct term, so that the use of the weak-charge approximation must remain doubtful.

Sorbello and Dasgupta (1977) have given a solution for the same model using quantum mechanics, but only to first order in the charge. Their qualitative conclusions agree with the results outlined above.

In drawing conclusions one must of course make the reservation that all the models that have been evaluated involve special and unrealistic assumptions. It seems reasonable to believe that in reality the force is not screened and that the only correction to the "direct" term is the wind effect. But further surprises cannot be ruled out.

The experimental situation on this problem does not seem too clear. The easiest quantity to measure is the mobility, but since this contains the resistance to the ion's motion it does not give direct information on the force. It should give the sign of the force, and it would be interesting to see whether, and in what cases, the ion moves in the "wrong" direction. In principle it is possible to observe the force directly by measuring the gradient in the ion density in diffusive equilibrium in an external field. I have not found any results of this type, and do not know whether the experiment is feasible.

4.4 Phonon Drag

We know that in the problem of heat conduction in clean insulating crystals the conservation of pseudomomentum plays an important part and is responsible for an exponential rise of the conductivity at low temperatures. (See *Surprises*, sec. 5.2). Does the same apply to the electric conductivity of pure metals?

Here we are concerned with the interaction between electrons and phonons. It will again be true that the total pseudomomentum is conserved, except for Umklapp processes, i.e., processes in which the pseudomomentum changes by $\hbar$ times a reciprocal-lattice vector, G. In the emission or absorption of a phonon by an electron, the process that is dominant in the resistance of a pure metal, we are concerned with

$$\mathbf{k}, \mathbf{f} \leftrightarrow \mathbf{k}', \tag{4.4.1}$$

where $\mathbf{k}$, $\mathbf{k}'$ are the wave vectors of the initial and final electron state and $\mathbf{f}$ that of the phonon. For an Umklapp process the conservation of pseudomomentum requires that

$$\mathbf{k} - \mathbf{k}' + \mathbf{f} = \mathbf{G}, \tag{4.4.2}$$

where $\mathbf{G}$ is again a reciprocal lattice vector. At low temperatures only phonons of small $\mathbf{f}$ are present, and the electron states involved are close to the Fermi surface. There can therefore be a substantial contribution from such processes only if there are points on the Fermi surface distant from each other by the smallest reciprocal-lattice vector.

For metals with a small number of conduction electrons, this condition is evidently not satisfied, and we would expect that for them the rate of Umklapp processes, and hence the resistivity, should decrease exponentially at low temperatures. At first sight this conclusion does not seem to apply to metals in which the conduction band contains few holes. The schematic figure 4.2a (drawn for a square lattice in two dimensions) shows that there are points on the Fermi curve distant by $\mathbf{G}$, for example the points labeled $\mathbf{k}$ and $\mathbf{k}'$. However, we can redefine the labeling of the wave vectors by counting them from one of the corners of the Brillouin zone, as shown in figure 4.2b. If these new labels are called $\hat{\mathbf{k}}$, they will also be subject to a selection rule of the form (4.4.2), so that at low temperature only processes with $\hat{\mathbf{G}} = 0$ can happen at an appreciable rate. We then again have an exact conservation law and therefore no resistivity.

On the face of it one would therefore expect the resistivity to decrease exponentially at low temperatures. However, one should re-

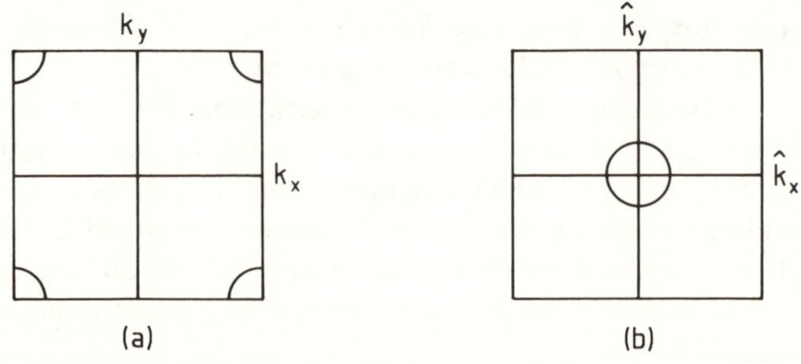

Fig. 4.2 Different labeling of hole states. (a) *Left*: counting wave vectors from zero. (b) *Right*: counting from zone corner.

member the existence of electron-electron collisions. These can also give Umklapp processes satisfying

$$\mathbf{k}_1 + \mathbf{k}_2 - \mathbf{k}_1' - \mathbf{k}_2' = \mathbf{G}. \tag{4.4.3}$$

This, with all four wave vectors on the Fermi surface, is much easier to satisfy than (4.4.2), and there are solutions for all cases except those with an extremely small number of electrons (or holes) in the conduction band.

So we would conclude that an exponential decrease of the resistivity should be quite rare, but that in many metals, including the alkalis, the resistance should depend essentially on the electron-electron collisions. Since these normally contribute very little to the resistivity, one expects a complication in the temperature dependence of the resistivity at low temperatures. However, the Bloch law, which is based on the assumption that the phonons are always in equilibrium, and which gives a T^5 law for the low-temperature resistivity, is very well obeyed.

I was aware of this puzzling contradiction in the 1930s, but wrote it down only later (Peierls 1955, sec. 6.7), and realized the explanation only much later.

It is difficult to obtain really pure specimens, and most measurements are made with impure metals that have a substantial residual resistance (resistance at $T = 0$). They are usually interpreted in terms of *Matthiessen's rule.* This says that the resistance of an impure specimen is the sum of the temperature-independent residual resistance, and that of an ideally pure metal. Until fairly recently there were no measurements in the low-temperature region in which the effects here discussed would have been important, and in which the ideal resistance was not a small correction to the residual resistance.

There is no basis for Matthiessen's rule. In particular, it is possible that impurities or lattice defects are effective in destroying the excess pseudomomentum of the phonons, while having a negligible effect on the electrons. In that case the resistivity should follow the law of Bloch, which assumes that the phonons are kept in equilibrium.

This was shown to be correct in experiments on samples of high purity, which showed deviations from Bloch's law at low temperatures. This was correctly interpreted as showing that the phonons could share in the forward movement of the electrons, and was verified in various other effects. For this reason the whole phenomenon is now referred to as *phonon drag*.

This is another problem that should not have surprised us if we had understood all the important factors.

5

Nuclear Physics

5.1 Rotational States

In early work on the structure of nuclei in the late thirties and forties, there was much speculation whether a nucleus could have rotational states like a solid body or a molecule. Some learned papers tried to prove that there could be no such states. The arguments used in these papers are too complicated to summarize here. There were, however, two reasons that had many of us believing there should be no rotational states. One was that experiments on nuclear excitation had not given any indication of such states. (We now know that well-defined rotational states are more common in heavier nuclei, whose large effective moment of inertia makes the rotational energies very small, so that these states show up only if high resolution is used.) The other reason was that atoms, which made up the other quantized many-body problem with which we were familiar, had no such rotational states—each electron could have an angular momentum, but there was no collective state of rotation.

It is indeed true that a spherically symmetric quantum system cannot rotate, because turning it through an angle is almost the same as exchanging the particles in it, and that is already contained in the ground-state wave function by its symmetry. Atoms have exact spherical symmetry only if they have closed electron shells, but even atoms with incomplete shells do not differ much from isotropy. The reason for this is that the interaction between the electrons is repulsive, and this causes them to distribute themselves as uniformly as possible.

In relying on the analogy between atoms and nuclei, we had not taken into account the vital difference that the interactions between nucleons are basically attractive. This causes nuclei with incomplete shells ("nonmagic numbers") to take nonspherical shapes, which allow nucleons to get closer to each other, without involving, by the Pauli principle, too much kinetic energy.

This idea was first invoked by Rainwater to explain the large quadrupole moments found in some nuclei. It was then realized, by Aage Bohr and others, that such a deformed nucleus could rotate. This must be thought of as a rotation of the surface, which represents the collective effect of all nucleons rather than of the matter inside. A double magic-number nucleus, with a spherical shape, has no rotational states associated with the ground state. We would expect the energies of the rotational states to be given by

$$E_L = E_0 + (\hbar^2/2I)L(L + 1), \tag{5.1.1}$$

where I is a moment of inertia and L the rotational quantum number, which runs through all integers if the nucleus has no mirror symmetry and only even values if there is mirror symmetry.

To calculate the moment of inertia, Inglis (1956) developed an ingenious method, now called the *cranking model*. He starts from a shell model with a deformed potential well and assumes the well to rotate with angular velocity ω. He then computes the energy increase of the nucleus, which should be of the form

$$E = \tfrac{1}{2}I\omega^2 + \text{const}, \tag{5.1.2}$$

so that

$$I = \frac{\partial^2 E}{\partial\omega^2}. \tag{5.1.3}$$

If the angular velocity ω is small enough, the energy can be calculated by perturbation theory. Taking for simplicity the case of rotation about a fixed axis, say the z axis, the perturbing term in the Hamiltonian is

$$V = \omega M_z = \hbar\omega L. \tag{5.1.4}$$

If the angular momentum in the ground state is zero, there is no first-order perturbation, and the standard expression for the second-order energy gives

$$\Delta E = \sum_{n \neq 0} \frac{|\langle 0|M_z|n\rangle|^2}{E_0 - E_n}. \tag{5.1.5}$$

Here we get our first surprise, because the quantity ΔE given by (5.1.5) is negative, since all E_n are greater than E_0.

The solution to this paradox is that ΔE is the change of energy in the co-moving frame, whose Hamiltonian is given by

$$H' = H + \omega M_z \tag{5.1.6}$$

if H is the Hamiltonian in the lab frame. The energy in the lab frame is what we require. Since H contains the rotating potential, it is not stationary and thus has no energy eigenvalues. We should therefore find its expectation value, which requires subtracting from H' the expectation value of ωM_z. But the expectation value of M_z in the perturbed state is given by $\partial H'/\partial\omega$, so that

$$\langle \Delta H \rangle = H' - 2\omega\frac{\partial E}{\partial \infty} = E_0 - \Delta E, \tag{5.1.7}$$

which represents correctly an increase in the energy.

However, there appears to be a more serious objection to the procedure. The potential well that we have used is only a calculational aid arising in the Hartree-Fock or similar procedure. We might expect, therefore, that if we applied the procedure to the exact states of a nucleus, the angular velocity would not appear, since it is defined only in connection with the potential well, and therefore the relation (5.1.3) would lead to $I = 0$.

For a long time this objection made me regard the cranking model as inconsistent. There is, however, a way out: note that the states we have used in the above argument are the eigenstates in the co-rotating frame, with the Hamiltonian H'. We have started from the eigenstates with $\omega = 0$, when H and H' are identical, and then applied perturbation theory. Now perturbation theory will always lead from the ground state of the unperturbed system to the ground state of the perturbed system. So we should look at the lowest eigenstate of H'.

Now assume we knew the exact eigenstates of H, which are also eigenstates of M. Moreover, assume that these form approximately a rotational band, so that near the ground state there will be a sequence of states with energies

$$H = E_0 + \frac{\hbar^2}{2I} L^2 \qquad (5.1.8)$$

and

$$M_z = \pm \hbar L. \qquad (5.1.9)$$

([5.1.8] differs from [5.1.1] since we are discussing rotation about a fixed axis.)

Using (5.1.6), (5.1.8), and (5.1.9), we plot the values of H' in figure 5.1 for the different states as a function of ω. The heavy line indicates in each case the lowest value of H'. If we now find the value of H in each case, we obtain figure 5.2, which is a discontinuous approximation to (5.1.2).

We see that the cranking-model procedure would make sense even if we knew the exact solutions, provided we always look for the lowest eigenvalue of H'. The surprise is that once again the "naive" intuitive approach is justified against a seemingly deeper analysis if this is not deep enough.

5.2 Projection Method

The cranking model described in the preceding section has limitations. For example, the wave function it generates is not an eigenstate of angular momentum, and that makes the discussion of selection rules difficult. It is therefore interesting also to consider alternative descriptions of rotational states. One of these starts from the variation principle. As explained in the preceding section, we expect nuclei with large incomplete shells to take a deformed shape, and that is confirmed by observations of their electric quadrupole moments. We therefore know that a Hartree-Fock type trial function would in these cases lead to an anisotropic potential. We shall talk about the deformed nucleus as elongated, but it would make no difference if it had an oblate shape.

But evidently the orientation in space of this elongated nucleus cannot matter, and therefore we have not one trial function, but a

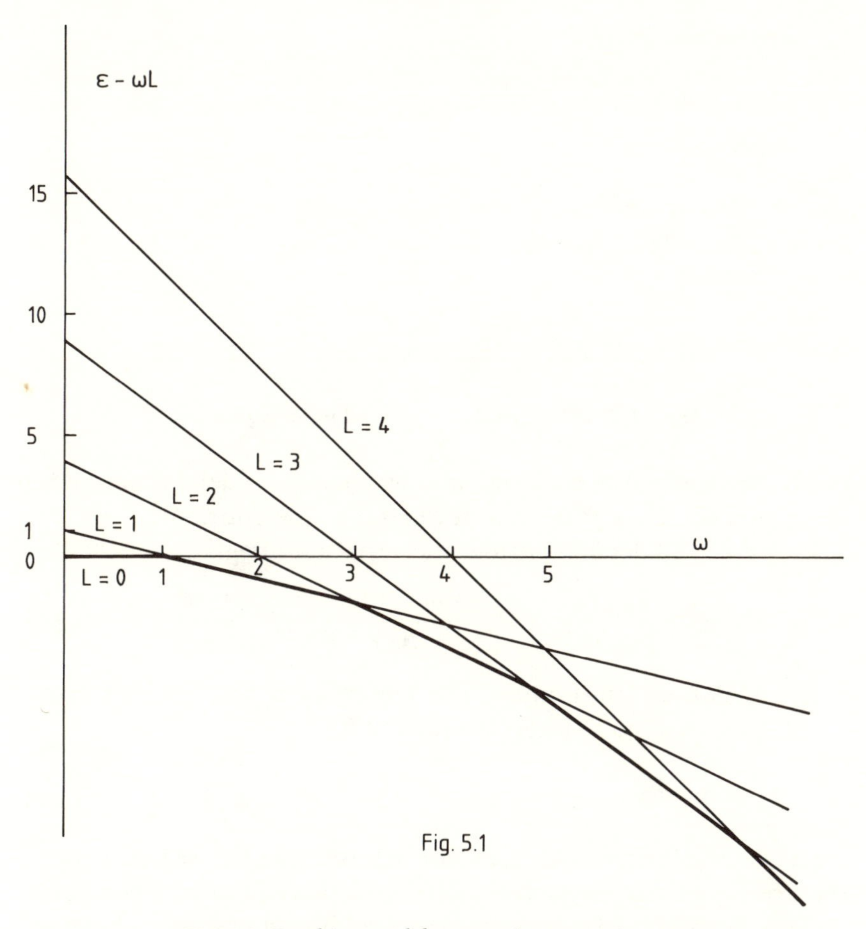

Fig. 5.1 Cranking model: energy in co-rotating system.

whole family, which I shall call ϕ_Θ, Θ indicating the orientation. I shall write out the arguments as if we were concerned only with rotation about a fixed axis; this will simplify the notation. We now have a kind of degeneracy, since all these functions give the same value to the variation integral. Now it is evident that we can further reduce the energy by choosing a linear combination of such functions,

$$\psi = \int d\Theta \, f(\Theta) \, \phi_\Theta, \tag{5.2.1}$$

with a suitable function f. It is clear that this will in general allow us to lower the variation integral, since the new expression (5.2.1)

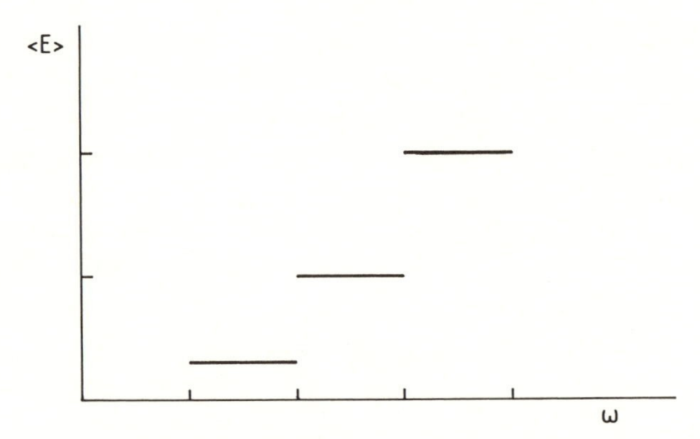

Fig. 5.2 Cranking model: energy values in lab system.

can be made equal to the original ϕ by choosing a delta function for f. But this choice of f will not in general be the optimum.

To see this in detail, consider the variation integral,

$$\langle E\rangle = \frac{\langle\psi|H|\psi\rangle}{\langle\psi|\psi\rangle} = \frac{\iint d\Theta\, d\Theta' f^*(\Theta)\, f(\Theta')\langle\phi_\Theta|H|\phi_\Theta'\rangle}{\iint d\Theta\, d\Theta' f^*(\Theta)\, f(\theta')\langle\phi_\Theta|\phi_{\Theta'}\rangle}, \tag{5.2.2}$$

where H is the Hamiltonian. The requirement that $\langle E\rangle$ be a minimum leads to a linear integral equation for f:

$$\int d\Theta' f(\Theta')\, \langle\phi_\Theta|H\phi_\Theta'\rangle - \langle E\rangle \int d\Theta' f(\Theta')\langle\phi_\Theta|\phi_{\Theta'}\rangle = 0. \tag{5.2.3}$$

Since the Hartree-Fock problem determining the ϕ is isotropic, the overlap matrix elements occurring in this equation depend only on $\Theta - \Theta'$, and the Θ integrations are folding operations. Therefore the solutions are of the form

$$f = \text{const}\ e^{im\Theta}, \tag{5.2.4}$$

with m an integer. In other words, ψ is now an eigenfunction of M_z, the appropriate component of angular momentum. If we had worked with a general orientation, ψ would become an eigenstate also of the resultant L.

To evaluate $\langle E\rangle$, and hence the moment of inertia, we have to compute the matrix elements occurring in (5.2.3). In general this is not simple, but in the most interesting cases, when there is an appreciable deformation involving many nucleons, i.e., far from closed shells, one expects the matrix elements to have a Gaussian dependence on $\Theta - \Theta'$, with the range of the Gaussian small compared to π, so that in the integrals $f(\Theta')$ may, for not too large m, be replaced by the first and second terms of a Taylor series for $f(\Theta')$ around Θ. In this approximation a simple calculation gives

$$E = E_0 + \frac{1}{2} m^2 \left[\frac{\{\langle\phi_0|H\phi_\Theta\rangle\}\,\{\Theta^2\langle\phi_0|\phi_\theta\rangle\}}{\{\langle\phi_0|\phi_\theta\rangle\}^2} - \frac{\{\Theta^2\langle\theta_0|H\phi_\theta\rangle\}}{\{\langle\phi_0|\phi_\theta\rangle\}} \right], \quad (5.2.5)$$

where the curly brackets denote integration over Θ. The expression in square brackets is then an estimate of the inverse of the moment of inertia, first used by Peierls and Yoccoz (1957) and still used by some authors.

The resulting estimates of the moment of inertia seemed reasonable, but here we had a very unpleasant surprise. The point is that the same method can be applied to another collective problem, namely the center-of-mass motion of the nucleus, if we replace Θ by the position of the center of the potential well, and the angular momentum m by the total wave vector k. This allows us to construct from a shell-model wave function an eigenfunction of the total momentum, and therefore eliminate, for example, the trouble described in *Surprises*, section 7.2. But in this case the dependence of the energy on the momentum k is known; it is just $k^2/2AM$, where A is the mass number and M the nucleon mass. The analog of (5.2.5) does not give anything like this result.

This finding is disastrous for the method, because if it gives the wrong answer to a problem for which the right answer is known, we cannot trust its application to a case in which the answer is not known.

What is the reason for this failure? The correct kinetic energy of a system follows from Galileo invariance, i.e., from the (nonrelativistic) rule for transforming to a moving observer. This says that if the wave function of a system is ψ, an observer moving with velocity v will ascribe to the system the wave function

$$\psi\, e^{(-im\mathbf{vR}/\hbar)}, \quad (5.2.6)$$

where $\mathbf{R}$ is the coordinate of the center of mass. From this it follows that the exact wave function must be of the form

$$e^{i\mathbf{kR}}\, \chi\,(\mathbf{x} - \mathbf{R}), \quad (5.2.7)$$

where χ depends, as indicated, only on the relative positions of the nucleons, and is independent of $\mathbf{k}$.

Now the analog of (5.2.1),

$$\psi = \int d\mathbf{r} e^{i\mathbf{k}\mathbf{r}} \phi(\mathbf{x} - \mathbf{r}), \quad (5.2.8)$$

is not of the form (5.2.7). We see this by forming $\psi \exp(-ik\mathbf{R})$, which should be independent of $\mathbf{k}$. With (5.2.8) this gives

$$\int d\mathbf{r} \; e^{i\mathbf{k}(\mathbf{r}-\mathbf{R})}\phi(\mathbf{x} - \mathbf{r} - \mathbf{R}), \quad (5.2.9)$$

which would be independent of $\mathbf{k}$ only if the exponent of the first factor were zero, i.e., if the center of mass coincided with the center of the potential well.

In other words, the family of functions defined by (5.2.1) is not Galileo invariant. One way of overcoming this trouble is to use the method of "minimizing after projection." In our use of (5.2.1) the function ϕ in the integrand was the Hartree-Fock eigenfunction, derived by minimizing the energy with a determinant wave function, and this was determined before we carried out the projection operation (5.2.1). If, instead, we leave the function ϕ undefined, and choose the function f as the angular-momentum projection, and then determine ϕ by minimizing the energy with the function (5.2.1), we preserve Galileo invariance, because the family of functions (5.2.1) with unspecified ϕ is Galileo invariant, in the sense that the Galileo transform of one function in the family is also a member of the family, though with a different ϕ. However, while this method is simple and elegant in idea, it leads to integrations that in practice are impossible to carry out. Practical applications are therefore possible only with further drastic approximations of dubious validity.

A better way of dealing with the problem in the case of center-of-mass motion is to repeat the original step. If we define (5.2.7) with a momentum $\mathbf{k}'$ and then perform a Galileo transformation that changes the momentum by $\mathbf{k} - \mathbf{k}'$, we have a new wave function of momentum $\mathbf{k}$ that is not identical with the first one. This means that we are again dealing with a degeneracy; we have again a family of functions with an equal right to be considered as good trial functions, and we can therefore improve the result by choosing a linear combination of them.

This leads to

$$\psi = \int d\mathbf{k}' d\mathbf{r}\, g(\mathbf{k}')\, e^{i(\mathbf{k}-\mathbf{k}')\mathbf{R} + i\mathbf{k}'\mathbf{r}}\, \phi(\mathbf{x}-\mathbf{r}), \tag{5.2.10}$$

which can be written as

$$\psi = e^{i\mathbf{kR}} \int d\mathbf{k}' d\mathbf{r}\, e^{i\mathbf{k}'(\mathbf{r}-\mathbf{R})}\, \phi(\mathbf{x}-\mathbf{r}). \tag{5.2.11}$$

In this form it is evidently of the form (5.2.7); the integral is a function of the relative positions only, since a displacement of all nucleons can be compensated by a change in the integration variable r.

With this trial function it is easy to evaluate the energy; with the Gaussian approximation for the overlap functions, we obtain an integral equation for g whose solution is again a Gaussian. The dependence of the energy on k is now evidently correct.

We cannot apply this idea directly to the case of rotation, because there is no analog of Galileo invariance (an angular velocity does alter the dynamics) and no analog of the center of mass. However, we can construct an analogy by noting that the function (5.2.10) is a linear combination of determinant functions with different centers and different velocities. In an analogous way Peierls and Thouless (1962) proposed a linear combination of functions with different orientations and different angular velocities.

Writing this again for the rotation about a fixed axis, we have

$$\int d\omega\, d\Theta\, g(\omega)\, e^{im\theta}\, \phi_{\Theta\omega}. \tag{5.2.12}$$

Here $\phi_{\Theta\omega}$ is the cranking-model wave function discussed in section 5.1, i.e., the Hartree-Fock type wave function minimizing $H + M_\omega$, with the deformed Hartree-Fock potential having orientation Θ. The weight function g is determined from the variation principle. Making again the Gaussian approximation for the overlap integrals, the solution for g also becomes a Gaussian.

The dependence of the energy on m determines a moment of inertia, and this turns out to be the same as that of the cranking model of section 5.1. For the three-dimensional rotation the same method applies in principle, but the calculations become more complicated, and no simple expression for the moment of inertia has been found.

5.3 Decaying States

One of the earliest applications of the Schrödinger equation was the theory of α decay by Gamow and by Condon and Gurney. They considered a problem with a potential barrier, which for the present we shall take as a rectangular barrier (fig. 5.3) to avoid mathematical complexity. In units in which $\hbar^2/2m = 1$, we have

$$\frac{d^2u}{dr^2} + (E - V)\,u = 0;\ u(0) = 0. \tag{5.3.1}$$

Here $u(r)$ is, as usual, the radial wave function $r\psi$. With the potential V shown in figure 5.3, there is evidently a continuous spectrum covering all positive energies.

However, if we impose the "outgoing boundary condition" that

$$u(r) \sim e^{ikr} \text{ as } r \to \infty, \tag{5.3.2}$$

where

$$k^2 = E, \tag{5.3.3}$$

it is easy to show that solutions exist only for complex eigenvalues, with energies

$$E_n = \epsilon_n - \tfrac{1}{2} i \Gamma_n, \tag{5.3.4}$$

with ϵ_n and Γ_n real.

For complex energy, the wave vector k will also be complex by

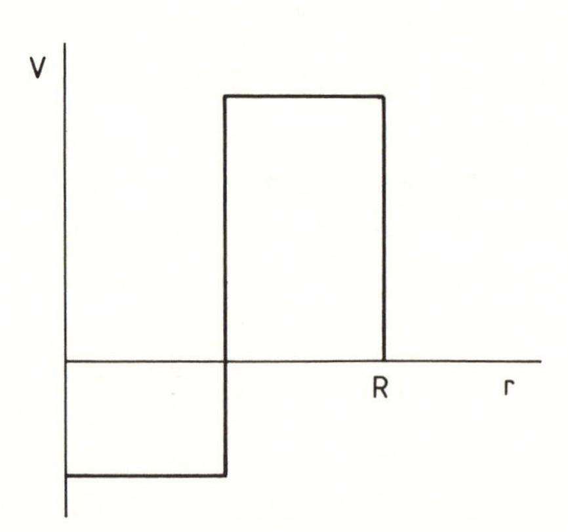

Fig. 5.3 Simple model with barrier penetration.

(5.3.3). In order for (5.3.2) to represent an outgoing flow, we choose the sign of k, which is not determined by (5.3.3), in such a way that the real part of k is positive. In that case it follows that the imaginary parts of the energy and of k are negative. To see this, multiply (5.3.1) by $u^*(r)$, the complex conjugate to u, subtract the complex conjugate of (5.3.1) times $u(r)$, and integrate from 0 to R, the distance beyond which the potential vanishes. After an obvious integration by parts, and noting that u and u^* vanish at $r = 0$, we find

$$\left(u^* \frac{\partial u}{\partial r} - u\frac{\partial u^*}{\partial r}\right)_R + (E - E^*) \int_0^R uu^* dr = 0. \qquad (5.3.5)$$

But since u must fit on smoothly to (5.3.2),

$$\frac{du}{dr} = iku \text{ at } r = R, \qquad (5.3.6)$$

the first term in (5.3.5) is equal to $(k + k^*)(uu^*)_R$. This shows that if the real part of k is positive, the imaginary part of E must be negative. We then see from the definition (5.3.3) that the imaginary part of k is also negative.

A negative imaginary part of E is obviously sensible. Since the time dependence of the state is given by $\exp(-iEt/\hbar)$, a negative imaginary part of E leads to an exponentially decreasing probability, in line with the fact that the probability of finding the particle inside the well decreases by tunneling through the barrier.

However, the negative imaginary part of k looks worrisome at first sight, and it was a surprise in the early days after the problem was first looked at. Indeed the wave function (5.3.2) now has a magnitude rising exponentially with r. This, however, is what we should have expected. The stationary wave equation implies that the state has existed forever, and, because of the exponential decay in time, that the wave was very much stronger in the past. At large distance we see the wave that originated from the center a long time ago and that reflects the more intense source at that time. It is easy to check this relation quantitatively, at least in the approximation that E and k are nearly real. The wave we see at time t at r started from small distances at $t' = t - kr$, since k is the velocity in our units. But at that time the amplitude was stronger by a factor $\exp(\Gamma r/2k)$, which by (5.3.3) is also $\exp|r.Imk|$.

However, as a consequence of the rise at large r, the wave function cannot be normalized, and it was sometimes thought that it was not a legitimate solution of the problem. This is true, in the sense that this function cannot by itself represent the real state of the system. This would also imply that the system started with infinite probability in the well. However, it is in this respect similar to a plane wave, which also cannot be normalized and therefore can never represent the state of a particle. But the plane wave is a very convenient tool; physical states can be obtained as superpositions of plane waves, and it would be most inconvenient if we could not use the idealization of a single plane wave for suitable purposes.

In particular, energies for which such decaying states exist show up as resonances in scattering. This is physically plausible because if a projectile strikes a target with the energy for which a decaying state exists, it may penetrate into the target and stay there for a time before emerging again. This is of course only a crude semiclassical description. It suggests that the scattering amplitude should have poles at complex energies corresponding to the complex eigenvalues we have discussed.

However, in following this idea we had another surprise. There are general theorems about the analytic properties of the "S matrix," whose connection with the scattering amplitude T is

$$S = 1 + 2ikT. \tag{5.3.7}$$

From an argument based on causality it can be shown that S, and hence T, has no singularities in the upper half of the complex energy plane. This would not trouble us because the eigenvalues in which we are interested have negative imaginary parts, so they are in the lower half. But other theorems that have to do with time-reversal symmetry and conservation of probability (unitarity) tell us that

$$S(E^*) = S^*(E), \tag{5.3.8a}$$

$$SS^* = 1, \tag{5.3.8b}$$

so there are no poles in the lower half-plane either. Where are the complex poles?

The answer to the paradox, which I reported to the 1954 International Conference on Nuclear Physics in Glasgow, was found by looking at explicit forms of the solution for simple scattering problems, but it can be seen very simply by considering again a one-body problem in a potential like figure 5.3. It is then known that the scattering amplitude is a regular function of k, except for poles, and that these poles are located roughly as shown in figure 5.4. Poles may exist on the positive imaginary axis, corresponding to bound states, with wave function decreasing at large distances. In addition, poles in the lower half-plane will be distributed symmetrically about the imaginary axis.

We can now describe the same situation in terms of energy. But because E is k^2, so that k and $-k$ belong to the same energy, the k plane covers the energy plane twice; in other words, we have to define our scattering amplitude on a Riemann surface. The two sheets of the Riemann surface are shown qualitatively in figure 5.5a,b. The one in 5.5a is usually called the *physical sheet*, because it contains the bound states at negative energy. The function has a branch point at $k = 0$, corresponding to the threshold. The branch point is the start of a cut, at which the two sheets are joined. This cut can otherwise be chosen arbitrarily. Scattering is usually discussed using the physical sheet, with the cut chosen along the positive real energy axis. In describing scattering for real positive energies, we must keep off the real axis, where the function is undefined, and one therefore adds an infinitesimal positive imaginary part, thereby keeping just above the real axis.

Now assume there is in the k plane a pole of the scattering amplitude T at

$$k_1 = \kappa_1 - \tfrac{1}{2}i\gamma_1 \tag{5.3.9}$$

with small γ_1. Then near the pole the leading term in T will be

$$T \approx A/k - k_1,$$

and because of (5.3.7), (5.3.8), a little algebra shows that, for small γ, the residue A has modulus $\tfrac{1}{2}\gamma_1$, so that

$$S \approx \frac{{}^1/_2\gamma_1 e^{i\alpha}}{k - k_l}, \tag{5.3.10}$$

where α is a phase which, for small γ_1, will be small.

Using the definition (5.3.3) this can be written as

$$S \approx \frac{{}^1/_2\gamma_1 e^{i\alpha}}{E - E_1}\,(k + k_1\,). \tag{5.3.11}$$

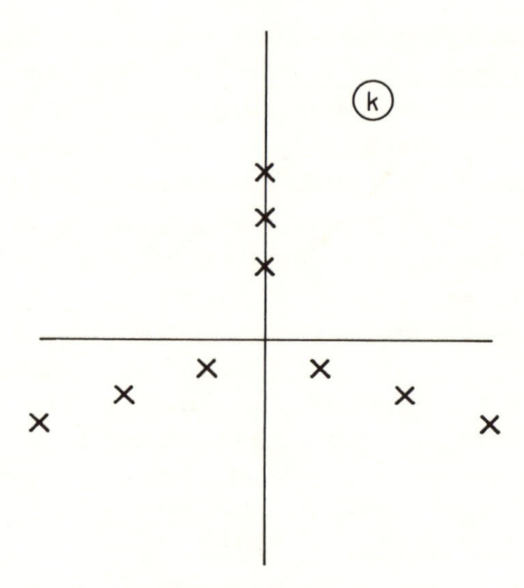

Fig. 5.4 Location of poles in k plane.

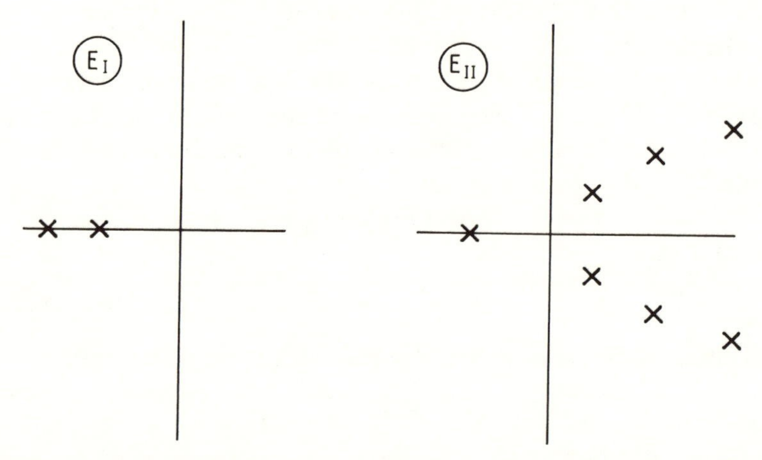

Fig. 5.5 Location of poles on Riemann surface for energy. (a) *Left*: "physical sheet." (b) *Right*: "unphysical" sheet.

In this form it is still explicit that the pole is only on one sheet of the Riemann surface, because at $E = E_1$ on the physical sheet, where $k = -k_1$, the pole cancels. The leading behavior is given by putting $k = k_1$ in the numerator, and then

$$S \approx \frac{{}^1\!/_2\,\Gamma_1 e^{i\alpha}}{E - \epsilon_1 - i\Gamma_1/2}, \tag{5.3.12}$$

and therefore the cross section

$$\sigma = 4\pi|T|^2 \approx \frac{4\pi|S|^2}{k^2} \approx \frac{\lambda^2}{\pi} \frac{\Gamma_1{}^2}{(E - \epsilon_1)^2 + \Gamma_1{}^{2/4}} \,. \qquad (5.3.13)$$

This is, for the one-channel case, the result of Breit and Wigner (1936). It shows a resonance at ϵ_1, of width Γ_1. The derivation seems to be simpler with k, and not E, as variable. However, the use of E and hence of a Riemann surface is essential when one considers the many-body problem, in which the target can have many states, so that there are thresholds at $E = \Delta_s$, where Δ_s is the sth excitation energy of the target. For an inelastic collision with excitation of the sth target level, the final wave vector is $\sqrt{(E - \Delta_s)}$, and the square root produces a branch point at every threshold. We can no longer flatten out the many sheets of the Riemann surface by using any k as variable.

In terms of energy, we can describe the effect of long-lived resonance states, i.e., of poles close to the real axis, by moving the (arbitrary) cuts down beyond the poles, thus justifying the analysis in terms of a Laurent series around each pole. Care has to be used in the case of poles close to threshold, since a cut is anchored at the branch point so that in pulling down the cut one cannot avoid complications.

Other complications arise when two or more poles are close together, with their distances comparable to their distances from the real axis (i.e., the energy differences comparable to the widths), because the Laurent series has a radius of convergence limited by the distance to the nearest singularity.

For this reason one cannot describe a situation with many resonances by the superposition of several Breit-Wigner type expressions, except in the trivial case that the resonances are so far apart that at any point only one of them matters. The early papers on the subject tended to rely on such a superposition, but its limitations are apparent if one notices that in situations where there is interference between several terms the result no longer satisfies unitarity.

A tempting idea is to try to obtain a rigorous formula from Cauchy's theorem, which involves integration over a contour. If this contour can be closed in the whole Riemann surface, it yields an expression for the amplitude in terms of all poles and their residues, with the added information that the residues can be deduced from the position of the poles (as they can in the one-body problem) and that there are identities involving the positions of the poles (Peierls 1959b). For the academic situation in which the force between the nucleons has a finite range beyond which it is sharply cut off, this proves possible in principle. However, in the more realistic case in which the asymptotic behavior of the forces is Yukawa-like, there appear new logarithmic branch points in the complex plane, which add an infinite number of sheets to the Riemann surface and make a complete analysis of the behavior on this surface impossible. This was another painful surprise.

More correct approaches to the problem are those by Kapur and Peierls (1938) (for a more modern version, see Peierls 1959a) and by Wigner and Eisenbud (1947). These developments are too complex to be summarized here, and they bring no particular surprises.

5.4 The Case of Overlapping Resonances

As the excitation energy of a nucleus increases, the density of resonance levels increases, and so does their width. Thus above some energy the width exceeds their spacing. This is the case of overlapping levels, sometimes rather misleadingly called the *continuous spectrum*. In his theory of the compound nucleus, Niels Bohr had given arguments to show that a nuclear reaction proceeded in two stages that are independent of each other: the formation of the compound nucleus by the merging of the incident particle with the target nucleus, and its decay. Therefore, the reaction cross section should be of the form

$$\sigma_{ab} = \sigma_{aC}\, \Gamma_b/\Gamma. \tag{5.4.1}$$

Here a, b stand for the different possible decay channels (defined by the nature of the emitted particles and the state of the residual nucleus, and their spins if appropriate), a being the incident channel. σ_{aC} is the cross section for the formation of a compound nucleus, Γ is the width of the compound state (so that $\Gamma/\hbar$ is its rate of decay), and Γ_b/h is the rate of decay into channel b.

The cross section for compound nucleus formation is, taking the case of an s-wave and neglecting spins for simplicity,

$$\sigma_{aC} = \frac{\lambda^2}{\pi} \Gamma_a/\Gamma, \tag{5.4.2}$$

which gives

$$\sigma_{ab} = \frac{\lambda^2}{\pi} \Gamma_a \Gamma_b/\Gamma^2, \tag{5.4.3}$$

which agrees with the Breit-Wigner result for many channels at full resonance, i.e., for $E = \epsilon_1$. Since in the case of overlapping levels one is effectively always in full resonance, it seemed to follow that the result (5.4.3) should apply.

But here was a paradox, because the result disagreed with measurements of the nuclear photoeffect in the region of overlapping levels, and also with a result derived from another general law, the law of detailed balancing of statistical mechanics. According to this very general law, if N nuclei are placed in an atmosphere of neutrons in thermal equilibrium with them, the rate of neutron capture must equal the rate of neutron emission.

Consider, for simplicity, a case on s-wave emission and absorption only, and neglect spins. Then the condition for balance is

$$N g_1 \Gamma_a/\hbar = N g_n v \sigma, \tag{5.4.4}$$

where g_1 is the statistical weight of the excitations, $\Gamma_a/\hbar$ the average decay rate of the excited states, g_n the statistical weight of the neutrons, v their velocity, and σ the average capture cross section. With our restrictive assumptions,

$$g_1 = E/d \; g_n = \frac{4\pi p^2 dp}{(2\pi\hbar)^3} = \frac{2\Delta E}{\hbar v \lambda^2}, \tag{5.4.5}$$

where d is the average spacing between relevant resonance states.

Hence (5.4.4) becomes

$$\sigma = {}^1/_2 \, \lambda^2 \Gamma_a/d \tag{5.4.6}$$

(λ is the neutron wavelength).

For well-separated levels this result agrees with the Breit-Wigner result (5.3.12), since the average cross section is here the sum of the energy integrals of the resonance cross sections for all levels, divided by the energy interval. However, in the case of overlapping levels it differs from (5.4.3) by a factor Γ/d. This paradox was resolved by Bohr, Peierls, and Placzek. Their argument was summarized in a letter to *Nature* (Bohr, Peierls, and Placzek 1939), but a promised detailed paper was never written (drafts are in Bohr 1986).

The solution depends on realizing that the meaning of the partial width Γ_a is different in (5.4.6) and in (5.4.3). In the case of overlapping levels, specifying the energy is not sufficient to know the state of the compound nucleus uniquely, since there are Γ/d different resonance levels whose energy equals the given value to within the uncertainty Γ. The state of the compound nucleus is then a superposition of all these levels with phases that may depend on the way the compound nucleus was formed.

In (5.4.3) we are talking about a specific state, namely that formed by capture from channel a. The partial width of this state for decay into channel a should for consistency be denoted by an extra superscript: $\Gamma_a^{(a)}$. In (5.4.4), on the other hand, we are concerned with statistical equilibrium, and that means a mixture of the relevant states. We shall denote the partial width for this mixture by $\Gamma_a^{(0)}$. The two need not be the same, and it is indeed plausible to expect $\Gamma_a^{(a)}$ to be greater, as this favors the configuration by which the state had been created.

We see therefore that both (5.4.3) and (5.4.6) are correct if the appropriate value of the partial width is used in each case. However, the quantity $\Gamma_a^{(0)}$ is usually more easily estimated from physical arguments, and therefore in practice the form (5.4.6) is to be preferred.

Note: Hauser and Feshbach (1952) derived a result equivalent to (5.4.6) by adding up the contributions from the Γ/d Breit-Wigner resonances within a width Γ, which seems a questionable procedure, though it gives the right answer (though without specifying the definition of the partial width).

5.5 Isospin of Nuclei

The early experiments on the force between nucleons showed that the "nuclear force" between two nucleons was the same regardless of whether one or both were neutrons or protons. This so-called "charge-independence" was at first an inspired guess, on the basis of measurements that were not really detailed enough to establish so general a theorem, but later was found to hold with considerable accuracy, though minor corrections remain. The theorem applies only to the specifically nuclear force, not including the electrostatic repulsion between two protons. The main experimental fact was the equality of behavior, apart from the Coulomb force, of the neutron-proton and the proton-proton system in the singlet-S state. The proton-proton system has no triplet-S state, because of the Pauli principle.

Collisions between two neutrons are not directly observable, but it was known from general experience in nuclear spectroscopy that there was *charge symmetry*, which meant that the behavior of nuclei is unchanged if all protons are replaced by neutrons and vice versa, again if allowance is made for the effect of the Coulomb forces. So proton-proton forces had to be equal (apart from Coulomb effects) to the neutron-neutron force.

Once this is established, it follows that there exists a new kind of symmetry in nuclei. We can see this in the following way. It is convenient to describe the neutron and the proton as two different states of the same particle, the nucleon, and introduce a variable capable of two values to distinguish its two states. This variable is then formally like a spin, and, just like a component, say σ_z can be $\pm 1/2$, we can introduce a variable for the nucleon, τ_3, which is $-1/2$ for the proton and $+1/2$ for the neutron. Since there exist exchange forces, under which nucleons change their nature, it is also convenient to have an operator that changes a neutron into a proton, which can be written as $2\tau_1$, where we are using spin matrices identical with the usual Pauli matrices:

$$\tau_1 = \begin{pmatrix} 0 & 1/2 \\ 1/2 & 0 \end{pmatrix} \quad \tau_2 = \begin{pmatrix} 0 & +1/2 i \\ +1/2 i & 0 \end{pmatrix} \quad \tau_3 = \begin{pmatrix} 1/2 & 0 \\ 0 & -1/2 \end{pmatrix} \qquad (5.5.1)$$

We shall use in particular the combination

$$\tau_+ = \tau_I + i\tau_2 = \begin{pmatrix} 0 & 0 \\ 1 & 0 \end{pmatrix}, \qquad (5.5.2)$$

which turns a neutron into a proton, and its conjugate, τ_-, which turns a proton into a neutron.

Now operate on the wave function of a nucleus with the sum

$$T_+ = \sum_s \tau_+^{(s)}, \tag{5.5.3}$$

where s labels the nucleon. The effect is to change all neutrons into protons. If the forces are charge-independent, this should not alter the effect of the potential on the wave function, so T_+ must commute with the Hamiltonian. Equally, T_- must commute with it. It follows that T_1 and T_2, the sums of the τ_1 and τ_2, respectively, for all nucleons, must commute with the Hamiltonian. The same goes for T_3, whose effect is just to multiply the wave function by one half the difference between the number of protons and the number of neutrons.

These matrices were first used by Heisenberg, who is sometimes given credit for inventing isospin. However, he used the matrices purely as calculational devices, without seeing the implication for symmetry. The symmetry argument is that we now have three matrices that satisfy the same algebra as the components of spin, and just as any quantity which commutes with angular momentum is isotropic in space, charge-independent interactions are isotropic on the abstract space spanned by our three matrices—isospin space.

As long as the assumption of charge-independence is justified, nuclear states can then be characterized by a given resultant isospin (just as states have given total angular momentum) and by a value of the component T_3, which measures the difference between neutrons and protons. So the components of an isospin multiplet are isobars, nuclei of the same mass number, which by these rules should have the same energy.

There also are selection rules, as with ordinary angular momentum: in the emission or absorption of a photon, the isospin can change at most by one unit, with $0 \rightarrow 0$ transitions forbidden for dipole radiation. In the emission and absorption of particles, the resultant isospin of the whole system must be conserved; so in the emission or absorption of a nucleon the isospin of the nucleus will change by $\pm \frac{1}{2}$, in the emission or absorption of an α particle, which has zero isospin, it remains unchanged, etc.

All these statements would be exact if there were exact charge independence. But we already noted that the Coulomb forces between protons violate charge independence. There are further small corrections from the mass difference between neutron and proton, plus other causes. Now the Coulomb force between two nucleons is considerably weaker than the nuclear force, and we might therefore expect that the isospin symmetry should be a reasonable approximation. Unfortunately, it differs from the nuclear force by being cumulative. The nuclear force saturates, and the binding energy of a nucleon in a heavy nucleus is of the same order of magnitude as in the α particle. On the other hand, the Coulomb force due to all other protons is additive, so that the mean electric potential acting on a proton in a heavy nucleus is Z times that in the α particle.

The effect of this is, in the first order, to shift each level by an amount depending on Z and to cause small corrections to the relative positions of the levels and to the selection rules. One expects that the isospin formalism would be useful only for the lightest nuclei.

Here, another interesting surprise came along. L. A. Radicati studied the effect of these corrections and found that the isospin formalism remained an excellent approximation even for medium-weight nuclei (Radicati 1953).

First of all, it is evident, he argued, that the neutron-proton mass difference does not mix states of different resultant isospin. This is because the nucleon charge is proportional to $(1\frac{1}{2} + \tau_3)$, thus the total effect is $(m_1 - m_2)$ $(\frac{1}{2} A + T_3)$, and a component commutes with the resultant spin. So the mass difference can only shift the different isobaric states but cannot affect the purity of states of given T.

If one adds a Coulomb interaction to a state of given isospin, say for the sake of the argument $T = 0$, this will cause admixtures to the wave function of terms belonging to a different isospin, say $T = 1$. However, these must belong to states with the same angular momentum and parity as the original state, and usually the lowest $T = 1$ state with the same angular momentum has a considerable excitation energy. Since the admixture depends on the ratio of the cross matrix element to the energy difference, this reduces the effect. Moreover, the cumulative part of the Coulomb energy, which is due to the effect on each proton of the mean Coulomb field of all the other protons, is a one-body effect, and therefore causes, in first order, transitions only to states differing, in the shell-model sense, in the state of one particle. It turns out that to find states of $T = 1$

that differ from the initial state by the transfer of one nucleon, one has to go to considerable excitations. Corrections to the shell-model picture will give other small corrections to this statement.

The Coulomb interaction between two nucleons outside closed shells can cause transfer of two nucleons, but these terms do not grow with increasing size of the nucleus.

Radicati estimated the corrections quantitatively and found, for example, that the admixtures of $T = 1$ states to $T = 0$ in nuclei with two valence nucleons, such as ^{6}Li or ^{14}C, is less than 1%, and in cases with four valence nucleons, less than 10^{-4}. It follows that isospin should remain a good quantum number to a reasonable approximation, well up to higher atomic weight, and this is borne out by experiment.

Once again we have got away with a cavalier approach, ignoring complications and believing in "naive" rules.

5.6 Isospin of Pions

When Yukawa put forward the meson hypothesis by which the force between nucleons is mediated by a meson field, whose quanta are now known as pions, there was clear evidence only of a nuclear force between unlike nucleons, i.e., between a proton and a neutron. This could be mediated by charged pions, and this would also produce exchange forces, which were then known to be at least part of the nuclear force.

However, arguments soon came up suggesting that there must also be a nuclear force between like nucleons, and this was definitely proved by experiments on proton-proton scattering. This suggested the existence of a neutral pion. But at first sight it did not seem easy to reconcile with the evidence for charge independence of the forces. The charged pion could, in first order, transmit a force only between unlike nucleons; for instance, a proton could emit a virtual π^+, becoming a neutron, and the π^+ could then be picked up by a neutron, turning it into a proton. On the other hand, a neutral pion could be emitted and absorbed by either kind of nucleon, so it would contribute to the interaction of both like and unlike nucleons. If the unlike-particle interaction was the sum of the two, and that between like particles only due to one effect, would not the one be necessarily stronger than the other, thus contradicting charge-independence?

The answer to this paradox turned out to be surprisingly simple, as follows from the work of N. Kemmer (1938). The answer is simply that the virtual exchange of a neutral pion between two unlike nucleons is repulsive and equal to half the strength of the contribution from charged pions. This answer sounds very contrived, but it follows automatically from a consideration of isospin.

Clearly the only simple way to make the nucleon-nucleon interaction isospin invariant is to make the pion-nucleon interaction invariant, i.e., to make the interaction an isoscalar. In first order the only possible form for this is

$$\text{const. } \sum_{\alpha} \tau_{\alpha} . \pi_{\alpha} , \tag{5.6.1}$$

using the τ_{α} defined in the preceding section, and denoting by π the wave function of the pion, which is an isovector with three components, corresponding to the three charge states. The suffix α runs over the isospin components.

When we apply this interaction to a second-order process in which a pion is exchanged between two nucleons, the result must preserve the symmetry and be an isoscalar. The only suitable form is the scalar product,

$$\text{const. } \tau^{a} . \tau^{b}, \tag{5.6.2}$$

where a and b refer to the two nucleons.

This expression is an isoscalar and therefore leads to charge independence. To see how this comes about, we can write the scalar product of (5.6.2) in the form

$$\tfrac{1}{2}(\tau_{+}^{a}\tau_{-}^{b} + \tau_{-}^{a}\tau_{+}^{b}) + \tau_{3}^{a}\tau_{3}^{b} , \tag{5.6.3}$$

using the definition (5.5.2). Labeling states in the order nn, np, pn, pp, we find by elementary algebra that the first two terms of (5.6.3) are

$$\tfrac{1}{2}\begin{pmatrix} 0 & 0 & 0 & 0 \\ 0 & 0 & 1 & 0 \\ 0 & 1 & 0 & 0 \\ 0 & 0 & 0 & 0 \end{pmatrix}, \tag{5.6.4}$$

which clearly corresponds, as expected, to an exchange interaction between neutron and proton, but does not contribute in the case of like particles. Similarly, the last term of (5.6.3) gives

$$\frac{1}{4}\begin{pmatrix} 1 & 0 & 0 & 0 \\ 0 & -1 & 0 & 0 \\ 0 & 0 & -1 & 0 \\ 0 & 0 & 0 & 1 \end{pmatrix}. \tag{5.6.5}$$

This is clearly the effect of the neutral pions, and it confirms that the action is opposite in the case of like particles. Adding the contributions together, we finally have

$$\begin{pmatrix} 1/4 & 0 & 0 & 0 \\ 0 & -1/4 & 1/2 & 0 \\ 0 & 1/2 & -1/4 & 0 \\ 0 & 0 & 0 & 1/4 \end{pmatrix}. \tag{5.6.6}$$

In the symmetric state of neutron and proton, which corresponds to an even orbital state with a spin singlet, the wave function of the $1/\sqrt{2}$ times the sum of the *np* and *pn* states, the coefficient is then $+1/4$, the same as for *nn* and *pp*. The antisymmetric state has no counterpart in the case of like particles, since by the Pauli principle it belongs to $T = 0$.

It is of course not necessary to go through this construction of matrix elements to get the answers; I spelled it out just to show how it comes about that the terms combine only to result in charge independence. In this case the use of the isospin concept can anticipate the surprise.

6

Field Theory

6.1 Classical Electron Models

H. A. Lorentz assumed that the electron has a finite radius in order to avoid the infinite self-energy of a point charge, even though there were difficulties in the relativistic description of an extended rigid object. He then expanded the equation of motion of a free electron, including radiative reaction, in powers of the electron radius r_0. We shall write this for low velocity, so that relativistic terms can be neglected, for simplicity.

$$m_m \ddot{x} = (A/r_0)\ddot{x} + B\dddot{x} + Cr_0\ddddot{x} + \dots \tag{6.1.1}$$

The appearance of derivatives of rising order of the position is required by dimensional considerations. Here m_{m} is the mechanical mass. The A term represents the electromagnetic self-energy, the B term is the well-known radiative damping, with

$$B = e^2/6\pi\epsilon_0 c^3. \tag{6.1.2}$$

The first term can be merged with the left-hand side to give mx, where

$$m = m_m + A/r_0 \tag{6.1.3}$$

is the total mass. The higher terms depend on the details of the charge distribution assumed.

Dirac, in a classical paper (1938), considered the equation in the limit $r_0 = 0$, but taking m, the total mass (6.1.3) to be finite. The resulting equation leads to reasonable consequences if the radiative reaction term is treated as a perturbation; but looking at exact solutions Dirac finds runaway solutions, of the form

$$x = \text{const.}\exp(6\pi\epsilon_0 mc^3/e^2)t, \tag{6.1.4}$$

in which the electron accelerates to infinite velocity (in a relativistic treatment to light velocity).

This surprising result can be understood if one remembers that by (6.1.3), if r_0 tends to zero for finite m, the mechanical mass m_{m} must tend to $-\infty$. As the electron accelerates, there is thus a large negative kinetic energy set free, which can balance the rising self-energy. For small accelerations the two infinities would cancel, leaving the effect of a finite total mass, but for high acceleration the damping term means that the self-field lags behind, so that the mechanical term wins.

I do not know whether Dirac would have agreed with this interpretation.

The difficulties arising from the infinities were avoided by the invention of renormalization of quantum field theory, which has made it possible to obtain results in spite of these infinities and has led to an impressive range of results in agreement with very accurate measurements. Yet renormalization theory is defined only in terms of a perturbation series. In the case of electromagnetism the series is probably only asymptotically convergent and thus does not strictly represent a function. However, the coupling constant $e^2/\hbar c$, which is essentially the expansion parameter, is so small that the doubts arising from the lack of convergence are negligible in any realistic calculation or experiment. For stronger interactions, however, this does not apply. Both for this reason, and because the use, as basic principle, of a semiconvergent series is unsatisfactory, there is still interest in theories that avoid singularities.

One such attempt was suggested by me and elaborated by McManus (1948). This led to classical (i.e., nonquantum) equations that seemed satisfactory, but the generalization to quantum field theory failed. It started from Lorentz's idea to regard the electron (and other charged particles) as having a finite extension, which would avoid the divergences coming in classical electrodynamics from the infinite self-energy of a point charge. However, special relativity does not allow a rigid object (because inside it the velocity of sound would be infinite and thus greater than light velocity), and so on the face of it an extended electron would require internal degrees of freedom, hence new fields, and probably new infinities. One way of avoiding this problem is the following.

An extended electron would imply that there would be charge at a space point different from the location of the electron. On computing the field, one would have to integrate over the space around the electron. But integration over space at constant time is not a relativistic concept. Why not replace it by an integration over space and time, which could preserve relativistic invariance? This implies replacing the relevant Maxwell equations by

$$\text{div}\,\mathbf{E} = \int d^3\mathbf{r}'dt' f(s^2)\,\rho(\mathbf{r}',t') \equiv \tilde{\rho} \tag{6.1.5}$$

$$\text{curl}\,\mathbf{H} - \dot{\mathbf{D}} = \int d^3\mathbf{r}'dt'\ f(s^2)\,\mathbf{j}(\mathbf{r}'.t') \equiv \tilde{\mathbf{j}}, \tag{6.1.5}$$

where, using units in which $c = 1$,

$$s^2 = (t - t')^2 - (r - r')^2 \tag{6.1.6}$$

is the invariant distance, and ρ and $\mathbf{j}$ are the density and flux due to the electron, in general δ functions. Similarly, the Lorentz force would be replaced by

$$\mathbf{F} = \int f(s^2)d^3\mathbf{r}'dt'[\mathbf{E}(\mathbf{r}',t') + \mathbf{v} \wedge \mathbf{H}(\mathbf{r}',t')], \tag{6.1.7}$$

where $\mathbf{r}$ is the position of the electron at time t, and $\mathbf{v}$ its velocity. Here we encounter a paradox. Since, for Lorentz invariance, the form factor f can depend only on the invariant distance s, it would seem that an electron would necessarily be strongly influenced by a field at a very distant point of that acted at a time at which a light signal would travel between the two points. This would not seem too disastrous if t' were prior to t, because then indeed a light signal could arrive at $\mathbf{r}$ at time t. But for consistency between (6.1.5) and (6.1.7), f must be symmetric between $\mathbf{r},t$ and $\mathbf{r}'t'$, so we would also expect advanced action, violating macroscopic causality.

Yet consider the Fourier tranform,

$$g(\mathbf{k},\omega) = \int d^3\mathbf{r}dt\ f(\mathbf{r},t)\exp i(\omega\, t - \mathbf{k}.\mathbf{r}). \tag{6.1.8}$$

If f is a function of the scalar s^2, g is a function of the scalar q^2:

$$q^2 = \omega^2 - k^2; \tag{6.1.9}$$

and if we choose g to be rapidly decreasing for both positive and negative values of the argument, this surely must cut out the effect of the high frequencies, but in the limit $r_0 = 0$, g will become a constant, so f becomes a four-dimensional δ function, hence the usual local theory, without any of the light-cone troubles.

This paradox is resolved by the properties of Fourier transforms in Lorentz space-time. If g is a smooth positive function decreasing for both positive and negative q^2, f is an odd function of s^2. Now consider for example the effective charge $\tilde{\rho}$ due to an electron distribution. Taking the effective charge density $\tilde{\rho}$ at $\mathbf{r} = 0$, $t = 0$, we have

$$\tilde{\rho}(0,0) = \int d^3\mathbf{r}dt\ \rho(\mathbf{r},t)f(s^2). \tag{6.1.10}$$

But for large $\mathbf{r}$, f is appreciable only near the light cone, $r = \pm t$. Taking the positive time as example, we can replace, in the neighborhood of the light cone, t by s as variable, $t = \sqrt{(s^2 + r^2)}$, and since f is appreciable only for s^2 of the order of $r_0{}^2$ or less, we can expand the square root in powers of s^2. If the electron motion is smooth, we can also expand in a Taylor series in t around $t = r$. This leads to

$$\tilde{\rho}(0,0) = \int d^3\mathbf{r} \left\{ M_0\rho(\mathbf{r},r) + M_l\left[\frac{1}{2r^2}\,\rho(\mathbf{r},r) - \frac{1}{2r}\,\frac{\partial\rho}{\partial t}\,(\mathbf{r},r)\right]\right.$$

$$\left. + M_2\left[\frac{3}{8r^4}\rho(\mathbf{r},r) - \frac{3}{8r^3}\,\frac{\partial\rho}{\partial t}\,(\mathbf{r},r) + \frac{1}{8r^2}\,\frac{\partial^2\rho}{\partial t^2}\,(\mathbf{r},r)\right] + \ldots\right\}, \tag{6.1.11}$$

where

$$M_n = \int ds^2\ s^{2n}f(s^2) \tag{6.1.12}$$

are the moments of the form factor. If g is even in q^2, f is odd in s^2, and all even moments vanish. Moreover, Chrétien and Peierls (1953) have shown that if the derivatives of g up to order N are bounded, then the moments of f up to order N vanish. In particular, if all derivatives of g are bounded, as for example for

$$g = \text{const.}\ \exp(-\alpha q^4), \tag{6.1.13}$$

the noncausal effects will decrease faster than any power of r, so there would be no macroscopic violations of causality.

Another surprise came when Irving (1949) applied the McManus formalism to the scattering of radiation by a free electron. For low frequency this of course gives the familiar classical Thomson formula. For high frequency we would have expected a reduced effect, but in fact there was an increase over the conventional classical result. The explanation of this surprise was interesting.

The scattering proceeds in three steps: (1) the force by the incident radiation on the electron; (2) the motion of the electron under the influence of this force; and (3) the emission of radiation by the oscillatory motion. In the McManus formalism, step (1) is unchanged, because the incident radiation has $q^2 = 0$, and its effect in (6.1.7) is modified by $g(0)$, which is 1. Similarly, step (3) is unchanged, because the emerging radiation must again have $q^2 = 0$, and therefore calculating the amplitude from the modified Maxwell equations (6.1.6) involves no change. What changes is step (2), because for the rapid oscillation the electron's self-field, and hence its inertia, is reduced. In fact, for the limit of high frequency the inertia should reduce to the mechanical mass only.

This feature might also appear in some more realistic theories. If we take the limit of r_0 going to zero, with constant total mass, so that the mechanical mass becomes negative, we find that for a rapidly accelerating motion the self-energy is reduced, thus backing the explanation of Dirac's runaway solutin that was proposed above.

It is not clear whether in this formalism there would be runaway solutions for a moderate r_0 when the mechanical mass is still positive.

This formalism is of course of no physical interest without generalizing it to quantum mechanics, and an attempt to do so failed (Chrétien and Peierls 1954).

6.2 Commutation Laws in Field Theory

The commutation laws of quantum field theory are usually expressed relating to two field quantities, say A and B, at different space points but at the same time, to get the commutator

$$[A(\mathbf{r},t),B(\mathbf{r}'t)]. \qquad (6.2.1)$$

Since the condition of equal times is not a Lorentz-invariant statement, this gives the rules a noncovariant form, although their content is of course invariant. Also, the usual rules are derived from a canonical formalism, which depends on the existence of a Hamiltonian (which requires that the equations of motion contain only the fields at one time and a finite number of their time derivatives), so it cannot be applied to formalisms (such as the one discussed in the preceding section) for which there exist a Lagrangian but no Hamiltonian.

It was generally regarded as impossible to formulate rules for the general commutator between A and B, where these might be field quantities at any space-time points, or, for example, integrals of field quantities over different space-time regions. The difficulty is that the variation of $[A(\mathbf{r},t),B(\mathbf{r}',t')]$ with t or t' is subject to the field equations, which cannot be explicitly solved. The surprisingly simple answer is to exploit this very fact to obtain a rule for the commutator. This goes as follows.

We assume that there exists a Lagrangian action principle, I. This determines the field quantities as a function of space-time if, for example, initial values at $t = -\infty$ are given. (We cannot write down the solution, but it exists in principle.) Now modify the action by an infinitesimal multiple of A,

$$I' = I + \lambda A, \tag{6.2.2}$$

with infinitesimal λ. The modified solution of the field equations will now differ from the original one by a term proportional to λ, and we write the retarded solution, i.e., the one that is not modified at large negative times, for any quantity B as

$$B' = B + \lambda D_{\mathrm{A}}B, \tag{6.2.3}$$

and the advanced solution, which is unmodified for long positive times, as

$$B'' = B + \text{ꓷ}_{\mathrm{A}}B. \tag{6.2.4}$$

With these definitions, a correct expression for the Poisson bracket is

$$\{A,B\} = (D_{\mathrm{A}}B - \text{ꓷ}_{\mathrm{A}}B). \tag{6.2.5}$$

It is evident that this satisfies the correct equations of motion. The two terms on the right-hand side are solutions of the same linear inhomogeneous equations, so their difference is a solution of the homogeneous equation for B. As regards the variation of A, adding a multiple of A to the action will not alter A, so (6.2.5) satisfies the correct equation for A.

In the case of an action that can be put in canonical form, one can show that the result is identical with the canonical Poisson brackets. In that case it is then also evidently true that the commutator

$$[A,B] = -i\hbar\{A,B\} = -i\hbar(D_A B - \mathfrak{a}_A B). \tag{6.2.6}$$

Note that it is not a priori evident that either the commutator or the Poisson bracket, as defined, is antisymmetric in A and B, as is necessary. However, it follows from the equivalence with the canonical forms, if the theory admits a canonical formalism.

The result (6.2.6) is useful in deriving relations in more convenient ways. For example, one can obtain the commutators for the electromagnetic field directly without introducing vector potentials, which are required for defining a Hamiltonian.

In principle, one can try to use the result as a definition of commutators in the case in which there exists an action principle, but no Hamiltonian (such as the formalism of the preceding section). This is possible evidently only if the required antisymmetry still holds. This can still be proved if A and B are fields or linear functions of them, but not for general operators. In general, the classical Poisson brackets as defined by (6.2.5) still have the required antisymmetry, but in the expressions for $[A,B]$ and $[B,A]$, noncommuting factors may appear in a different order, and there is no guarantee that the terms left by switching them around will cancel. However, for most purposes knowing the commutators of the fields should be adequate.

The method can also be extended to fermion fields, by defining a quantity Θ which anticommutes with all fermion fields, and allowing only Θ times a fermion field as an addition to the action.

For details see Peierls (1952).

7

Hydrodynamics

7.1 Drag on a Sphere

The difficulty that arose in section 2.5 is reminiscent of a situation in a very old and familiar problem, though the surprising aspect is not so well known.

This is the problem of the irrotational flow of an ideal incompressible liquid surrounding an immersed sphere of radius a, which moves with velocity U. The well-known solution is derived from the velocity potential,

$$\phi = -Ua^3z/2r^3, \tag{7.1.1}$$

where r is the distance from the center and z its component in the direction of motion.

In this idealized picture there is no drag on the sphere in uniform motion; but if the sphere is accelerating, there is a drag, which represents the force necessary to accelerate the liquid in the neighborhood. There are two easy methods to compute this drag. One is to calculate the kinetic energy of the motion described in (7.1.1) and then equate its rate of change with the rate of work done by the sphere against the drag F:

$$F.U = \frac{d}{dt} E_{\mathrm{kin}}. \tag{7.1.2}$$

The kinetic energy is from (7.1.1):

$$E_{\mathrm{kin}} = \frac{2\pi}{3}\rho a^3 U^2, \tag{7.1.3}$$

and therefore from (7.1.2):

$$F = \frac{4\pi}{\partial}\rho a^3 \dot{U}. \tag{7.1.4}$$

The same result can also be derived by computing the pressure on every element of the surface of the sphere and adding up the resultant.

However, the physically most obvious approach would be to determine the rate of change of momentum of the liquid, which should also equal the drag. This requires an expression for the liquid momentum in the z direction:

$$p_z = \frac{1}{2}\rho U a^3 \int d^3\mathbf{r}\, \frac{3z^2 - r^2}{r^5}. \qquad (7.1.5)$$

This is the same type of ambiguous integral we encountered in the problem of the momentum of a sound wave (sec. 2.5). However, in this case we know the correct answer, (7.1.4), and can ask what shape of the integration volume will give this answer. It turns out that the right shape is a narrow, elongated volume in the z direction.

In this case it is not necessary to choose the correct shape by working back from the answer. We can also apply the law of conservation of momentum to a finite volume, before going to the limit. It takes the form

$$\frac{d}{dt}\int_V p_i d^3\mathbf{r} = \rho\int_S d\sigma\, u_n u_i - \int_S d\sigma\, P n_i. \qquad (7.1.6)$$

Here V is any volume, S its boundary surface; $d\sigma$ is an element of area on the surface; u_n is the component of the velocity normal to the surface; and n_i is the component of the unit normal vector in the direction i. P is the pressure.

We need the momentum in the z direction, so $i = z$. If we choose for V a cylinder of length $2L$ and radius A, with its axis in the z direction and its center at the origin, then the first term on the right-hand side of (7.1.6) has no contribution from the end faces, because on the two faces the velocity is the same, but u_n has opposite signs on the two faces. The contribution from the curved surface is non-zero; but it decreases as the inverse fourth power of the linear dimensions, and so is negligible in the limit.

There is, however, a contribution from the second term. The pressure is

$$P = \rho\,\dot{\phi} - {}^1/_2\rho u^2 + \text{const.} \qquad (7.1.7)$$

Since in (7.1.6) P is multiplied by n_z, only the end faces contribute, and the last two terms in (7.1.7) have the same value on the two end faces and thus cancel. The remaining term can be evaluated from (7.1.1) and gives

$$\pi\rho\dot{U}\left[1 - \frac{L}{(L^2 + A^2)^{1/2}}\right]. \tag{7.1.8}$$

It is now obvious that for a very elongated cylinder, $L \gg A$, this expression vanishes in the limit, whereas in the case of any other shape a finite surface correction remains. Therefore we may invoke conservation of momentum in the case of the elongated shape without correction, but not otherwise. Thus we can fully understand the origin of our surprise in this case; but the argument is not immediately applicable to the problem of section 2.5, since we are there concerned with a total momentum, not with a force.

7.2 Shock Waves

An elegant mathematical approach to one-dimensional flow, including shock waves, was given by Riemann (1901). Consider the one-dimensional flow of a compressible but nonviscous fluid. The Euler equation of motion is then

$$\frac{\partial u}{\partial t} + u\frac{\partial u}{\partial x} + \frac{c^2}{\rho}\frac{\partial p}{\partial x} = 0, \tag{7.2.1}$$

where u is the velocity, ρ the density, and $c^2 = dp/d$ the square of the sound velocity, with p the pressure.

The continuity equation is

$$\frac{\partial \rho}{\partial t} + u\frac{\partial \rho}{\partial x} + \rho\frac{\partial u}{\partial x} = 0. \tag{7.2.2}$$

These two equations are nonlinear partial differential equations, and one cannot easily seem to make sense of them. Riemann found a transformation which, by an elegant use of the concept of characteristics, led to much useful information. By introducing an auxiliary variable σ, a function of the density,

$$\sigma = \int \frac{c}{\rho}\, d\rho \tag{7.2.3}$$

$$\left\{\frac{\partial}{\partial t} + (u \pm c)\frac{\partial}{\partial x}\right\}(\sigma \pm u) = 0. \tag{7.2.4}$$

This shows that at a point moving in the $+x$ direction with velocity $c + u$, the combination $\sigma + u$ is constant, and at a point moving toward $-x$ with speed $c - u$ the quantity $\sigma - u$ is constant. In general, both c and u vary, so the evaluation of these statements is not easy, though this form of the equations is more convenient for numerical work than the original form.

The power of the method becomes clear in the case in which a disturbance, caused, for example, by a moving piston, moves into fluid at rest at uniform density. Figure 7.1 shows an x, t diagram, with the heavy line the motion of the piston. We have drawn lines of points moving with speed $c + u$ forward, or $c - u$ backward, respectively, and we shall refer to these as the forward or backward characteristics. Consider first a backward one, such as BB'. It extends into the region of undisturbed fluid, where $\sigma - u$ is equal to σ_0, the value of σ at the original density. So all along such a line $\sigma - u = \sigma_0$. But these backward characteristics fill the whole plane, so this identity applies throughout.

Now consider a forward characteristic, such as AA'. All along it

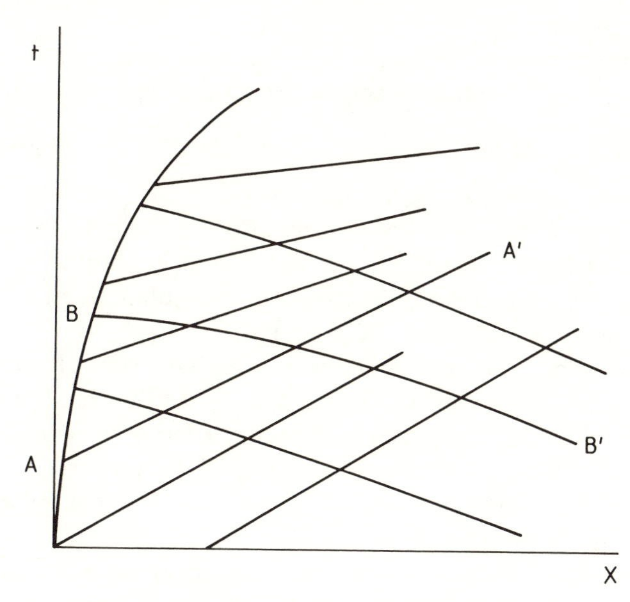

Fig. 7.1 One-dimensional compressible flow.

$\sigma + u$ is constant, but since $\sigma - u$ is also constant, it follows that along AA' both u and σ, and hence c, are constant. So the forward characteristics are straight lines, as we have drawn them, though each with a different slope. If the piston is accelerating, as drawn, the values of u and c increase. The first follows from the fact that the velocity must match that of the piston, the other then follows from the constancy of $\sigma - u$. Evidently the forward characteristics will cross, which does not seem to make sense, because some quantities must then have different values at the same point. Riemann recognized that such a crossing implied a discontinuity, though he did not discuss what happens beyond the crossing point.

He also realized that the boundary conditions at such a discontinuity must be obtained from the conservation laws. These are most easily written down in a coordinate system in which the shock is stationary. If the shock velocity is S, we then have a fluid entering with velocity S from the right and leaving with velocity $S - u$ to the left. Conservation of mass says that the quantity of fluid passing per unit time and per unit area must be the same, i.e.,

$$S \rho_1 = (S - u)\rho_2, \qquad (7.2.5)$$

where the subscripts 1,2 relate to the conditions ahead of and behind the shock.

Conservation of momentum says that the difference in the fluid momentum must equal the thrust due to the pressure difference:

$$p_2 - p_1 = \rho_1 S^2 - \rho_2 (S - u)^2 = \rho_1 S u = \rho_2 (S - u) u. \qquad (7.2.6)$$

(For the last two equalities we have used the first relation [7.2.5].)

If one regards the pressure as a function of the density, there are three quantities, S, u, and ρ_2 in these two equations, so shocks would be a one-parameter family, which is plausible because the strength of the shock has to be a parameter.

However, there is a further conservation law, that of energy. Riemann was aware of this and found that he had no further freedom to satisfy this, and here was a paradox. Riemann even suggested that in the shock some energy might be lost! Others proposed to base the calculations on conservation of mass and energy, but then of course momentum was not conserved. This difficulty led to the whole subject acquiring a dubious reputation among mathematicians. In the updated version of the "Riemann-Weber," by Frank and von Mises, which still retains the names Riemann and Weber in the subtitle, the subject of the two chapters on one-dimensional flow and shock waves is not mentioned. At the same time experimentalists, such as Mach, continued to do beautiful work on shock waves.

The fact that Riemann overlooked is that the state of a fluid is determined by two independent variables, density and temperature, or density and entropy. If the time scale of the flow is short and one can therefore neglect viscosity and heat conduction, then the flow will be adiabatic, i.e., the entropy is constant, and the variation of p with ρ assumed in our basic equations has to be the adiabatic relation.

The fact that on this basis we cannot match conservation of energy indicates that the shock is an irreversible process, in which entropy increases. This also explains why the construction of figure 7.1 is not applicable beyond the shock front.

Energy conservation is expressed in the following way:

$$\epsilon_2 - \epsilon_1 + P_2 v_2 - p_1 v_1 + {}^1/_2(S - u)^2 - {}^1/_2 S^2 - 0, \quad (7.2.7)$$

where ϵ is the internal energy per unit mass, and $v = 1/\rho$ the specific volume. Expressing the pressures in terms of the mean pressure $\bar{p} = \frac{1}{2}(p_1 + p_2)$ and the difference (7.2.6), a little algebra leads to

$$\epsilon_2 - \epsilon_1 = \bar{p}(v_1 - v_2). \quad (7.2.8)$$

In this form the incompatibility with the adiabatic condition is clear: the relation (7.2.8) holds good for infinitesimal differences, because it then says $d\epsilon/dp = -v$, which is indeed true in an adiabatic equation of state. However, for finite difference the relation is not the same (unless the relation between pressure and volume were linear).

The microscopic theory of shock waves in a gas was given by Lord Rayleigh (1910) and by G. I. Taylor (1910). It shows that in reality the change in pressure and density is not sudden, but is spread over a distance of the order of the mean free path. In that region viscosity and heat conduction are no longer negligible, and this accounts for the irreversible nature of the shock.

References

Amos, A. T. 1970. *Chem. Phys. Lett.* 5:587.

Blount, F. I. 1971. Unpublished report.

Bohr, N. 1986. *Collected Works*, vol. 9, pp. 487, 503. North Holland.

Bohr, N.; Peierls, R.; and Placzek, G. 1939. *Nature* 144:200.

Breit, G., and Wigner, E. P. 1936. *Phys. Rev.* 49:519, 642.

Burt, M. G., and Peierls, R. 1973. *Proc. Roy. Soc.* A33:149.

Chrétien, M., and Peierls, R. 1953. *N. Cimento* 10:668.

———. 1954. *Proc. Roy. Soc.* A223:465.

Das, A. K., and Peierls, R. 1973. *J. Phys.* C6:281.

———. 1975. *J. Phys.* C8:3348.

Dirac, P.A.M. 1938. *Proc. Roy. Soc.* A167:148.

Eisenschitz, R., and London, F. 1930. *Z. Physik* 60:491.

Gordon, J. 1973. *Phys. Rev.* A8:14.

Gorter, C. J., and Casimir, H. 1934. *Physica* 1:306.

Hauser, W., and Feshbach, M. 1952. *Phys. Rev.* 87:366.

Hirschfelder, J. O. 1967. *Chem. Phys. Lett.* 1:325.

Inglis, D. R. 1956. *Phys. Rev.* 103:1786.

Irving, J. 1949. *Proc. Phys. Soc.* 42:780.

Ising, E. 1925. *Z. Physik* 31:253.

Jones, H. 1934a. *Proc. Roy. Soc.* A144:225.

———. 1935b. *Proc. Roy. Soc.* A147:396.

Jones, R. V., and Leslie, B. 1978. *Proc. Roy. Soc.* A360:347.

Jones, R. V., and Richards, J.C.S. 1954. *Proc. Roy. Soc.* A221:480.

Kapur, P. L. 1937. *Proc. Roy. Soc.* A163:353.

Kapur, P. L., and Peierls, R. 1937. *Proc. Roy. Soc.* A163:606.

———. 1938. *Proc. Roy. Soc.* A166:277.

Kemmer, N. 1938. *Proc. Camb. Phil. Soc.* 34:354.

Lal, H. M.; Suen, W. M.; and Young, K. 1981. *Phys. Rev. Lett.* 47:177.

Landau, L. D., 1937. *Phys. Z. der Sovjetunion* 11:129.

Landau, L. D., and Lifshitz, E. M. 1960. *Electrodynamics of Continuous Media*. Pergamon.

Landauer, R. 1975. *J. Phys.* C8:761.

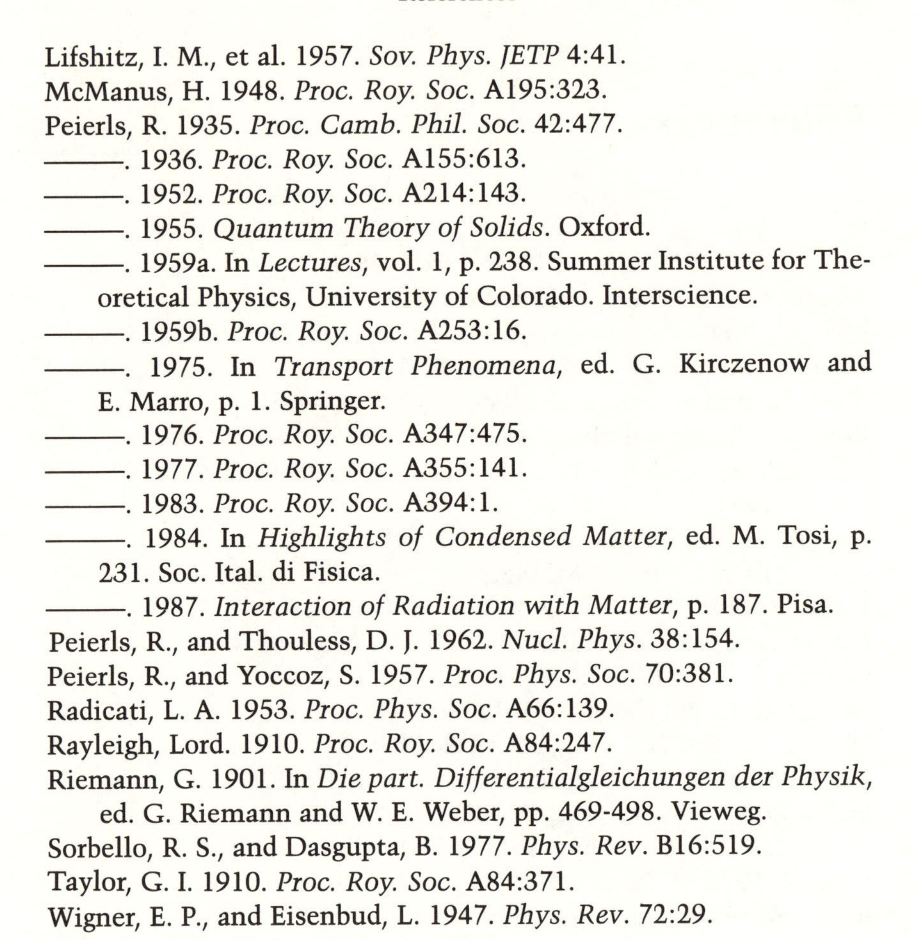

Lifshitz, I. M., et al. 1957. *Sov. Phys. JETP* 4:41.
McManus, H. 1948. *Proc. Roy. Soc.* A195:323.
Peierls, R. 1935. *Proc. Camb. Phil. Soc.* 42:477.
———. 1936. *Proc. Roy. Soc.* A155:613.
———. 1952. *Proc. Roy. Soc.* A214:143.
———. 1955. *Quantum Theory of Solids*. Oxford.
———. 1959a. In *Lectures*, vol. 1, p. 238. Summer Institute for Theoretical Physics, University of Colorado. Interscience.
———. 1959b. *Proc. Roy. Soc.* A253:16.
———. 1975. In *Transport Phenomena*, ed. G. Kirczenow and E. Marro, p. 1. Springer.
———. 1976. *Proc. Roy. Soc.* A347:475.
———. 1977. *Proc. Roy. Soc.* A355:141.
———. 1983. *Proc. Roy. Soc.* A394:1.
———. 1984. In *Highlights of Condensed Matter*, ed. M. Tosi, p. 231. Soc. Ital. di Fisica.
———. 1987. *Interaction of Radiation with Matter*, p. 187. Pisa.
Peierls, R., and Thouless, D. J. 1962. *Nucl. Phys.* 38:154.
Peierls, R., and Yoccoz, S. 1957. *Proc. Phys. Soc.* 70:381.
Radicati, L. A. 1953. *Proc. Phys. Soc.* A66:139.
Rayleigh, Lord. 1910. *Proc. Roy. Soc.* A84:247.
Riemann, G. 1901. In *Die part. Differentialgleichungen der Physik*, ed. G. Riemann and W. E. Weber, pp. 469-498. Vieweg.
Sorbello, R. S., and Dasgupta, B. 1977. *Phys. Rev.* B16:519.
Taylor, G. I. 1910. *Proc. Roy. Soc.* A84:371.
Wigner, E. P., and Eisenbud, L. 1947. *Phys. Rev.* 72:29.